Geography

Polity's *Why It Matters* series

In these short and lively books, world-leading thinkers make the case for the importance of their subjects and aim to inspire a new generation of students.

Lynn Hunt, *History*
Tim Ingold, *Anthropology*
Neville Morley, *Classics*
Alexander B. Murphy, *Geography*
Geoffrey K. Pullum, *Linguistics*

Alexander B. Murphy

Geography

Why It Matters

polity

First published in 2018 by Polity Press
Reprinted 2019 (twice), 2020, 2021

Polity Press
65 Bridge Street
Cambridge CB2 1UR, UK

Polity Press
101 Station Landing
Suite 300
Medford, MA 02155, USA

ISBN-13: 978-1-5095-2300-9
ISBN-13: 978-1-5095-2301-6 (pb)

A catalogue record for this book is available from the British Library.

Typeset in 11 on 15 Sabon by Servis Filmsetting Ltd, Stockport, Cheshire
Printed and bound in the United States by LSC Communications

The publisher has used its best endeavours to ensure that the URLs for
external websites referred to in this book are correct and active at the time
of going to press. However, the publisher has no responsibility for the
websites and can make no guarantee that a site will remain live or that the
content is or will remain appropriate.

Every effort has been made to trace all copyright holders, but if any have
been inadvertently overlooked the publisher will be pleased to include any
necessary credits in any subsequent reprint or edition.

For further information on Polity, visit our website: politybooks.com

For

R. Taggart Murphy

my brother, life-long supporter, and friend,
who has always encouraged me to think more deeply
about the world around us

Contents

Acknowledgments

The ideas set forth in this book first took root when I chaired a US National Research Council study outlining strategic directions for the geographical sciences. I am grateful to the other participants in that study for the many ideas and insights that made their way into the study report and into this book. A residency fellowship several years ago at the Rockefeller Foundation's Bellagio Center gave me the opportunity to develop my thinking about geography's significance, and a return engagement in the summer of 2017 gave me the time to put together a first draft of this manuscript. On both occasions I benefited from the ideas and insights of my fellow Bellagio residents. It is no exaggeration to say that support from the Rockefeller Foundation was crucial to the completion of the project.

Acknowledgments

I am also grateful to many geography colleagues and students at the University of Oregon and beyond who shared ideas and suggestions. I am particularly indebted to Patrick Bartlein, Eve Vogel, Mark Fonstad, Jerilynn "M" Jackson, Daniel Gavin, Leslie McLees, Anna Moore, Craig Colton, Diana Liverman, Carlos Nobre, and David Kaplan. Dean Olson served as my research assistant as I finished up the manuscript; his input and help with research and figures were invaluable. I also thank my sister, Caroline Murphy, and my brother (to whom the book is dedicated) for insightful feedback on portions of the book.

The manuscript benefited greatly from the comments of two anonymous reviewers and from the helpful input of my editor, Pascal Porcheron. Finally, Susan Gary has been at my side throughout my work on this book and so much else through the years. My debt to her is beyond words.

Illustration Credits

Mid-Holocene Climate," *Geophysical Research Letters*, 44:17 (2017): 9022.

3 Figure originally created by Alexander B. Murphy and Nancy Leeper for *Geographical Approaches to Democratization: A Report to the National Science Foundation* (printed by the University of Oregon Press for the Geography and Regional Science Program, National Science Foundation, 1995).

4 Reproduced with permission from D. J. Weiss, A. Nelson, H. S. Gibson, W. Temperley, S. Peedell, A. Lieber, M. Hancher, E. Poyart, S. Belichior, N. Fullman, B. Mappin, U. Dalrymple, J. Rozier, T. C. D. Lucas, R. E. Howes, L. S. Tusting, S. Y. Kang, E. Cameron, D. Bisanzio, K. E. Battle, S. Bhatt, and P. W. Gething, "A Global Map of Travel Time to Cities to Assess Inequalities in Accessibility in 2015," *Nature*, 533 (2018): 334.

5 Modified from Richard Edes Harrison, *Fortune Atlas for World Strategy* (New York: Alfred A. Knopf, 1944), pp. 8–9.

6 Modified from Kai Krause, *The True Size of Africa* (2010). Available at http://kai.sub.blue/en/africa.html.

7 Reproduced with permission from K. O'Brien, R. Leichenko, U. Kelkar, H. Venema, G. Aandahl, H. Tompkins, A. Javed, S. Bhadwal, S. Barg, L. Nygaard, and J. West, "Mapping Vulnerability to Multiple Stressors: Climate Change and Globalization in India," *Global Environmental Change*, 14:4 (2004): 307.

Illustration Credits

8 Map by Derek Watkins that appeared in conjunction with Andrew E. Kramer and Andrew Higgins, "Ukraine's Forces Escalate Attacks Against Protesters" (*New York Times International*, February 21), pp. A-1 and 11. Reproduced with permission, © New York Times.

9 Reproduced with permission from Matthew J. Kauffman, James E. Meacham, Alethea Y. Steingisser, William J. Rudd, and Emiliene Ostlind, *Wild Migrations: Atlas of Wyoming's Ungulates* (Corvallis, OR: Oregon State University Press, 2018), p. 139. © 2018 University of Wyoming and University of Oregon.

1

Geography's Nature and Perspectives

Imagine that you could transport yourself back in time to the early 1960s and visit the Lake Chad region in Africa. You would find yourself wandering along the shores of one of the largest lakes in Africa – a lake straddling the boundaries of four newly independent countries: Chad, Cameroon, Nigeria, and Niger. You would see a rich lake-centered ecosystem providing life-giving water and food for a few million people living near its shores. Most of the people you met would rely on the lake's abundant fish harvests, but you would notice farming and pastoralist communities as well. You might hear stories of tensions between different ethno-cultural groups, but not of armed conflict. During your explorations of the physical environment near the lake you would find significant woodland stands in some places, but sparser vegetation elsewhere

1

because of the challenges presented by the long winter dry season. You might well be aware of the ecological fragility of the region, but you would be encouraged by an agreement entered into by the four states sharing the Lake Chad Basin setting forth a plan for cooperative management of the Basin's development.

A visit today would be a very different experience. You would find a lake that has lost 90 percent of its 1960 surface area (plate 1), and a fish population that is a shadow of its former self. You would see a human population more than twice its 1960s size, but with abandoned villages in some areas and newly established makeshift settlements in others. As you wandered around you would encounter far fewer people making their living from fishing and far more from agriculture, and you would see evidence of major land-use conflicts resulting from the expansion of agriculture into pasture land. You would also likely be aware of deep tensions between the different states controlling parts of the Basin – and indeed between state authorities and local peoples.

You would also see the impact of Boko Haram, a radical Jihadist insurgency movement that took root in northern Nigeria in the early 2000s and instigated an armed uprising aimed at establishing an Islamic state grounded on strict (many would say corrupt) Sharia law principles. Boko Haram's advance into the Lake Chad region, and the military response of autocratic governmental authorities (often with support from the West), resulted in the displacement of well over two million people, the loss of thousands of lives to conflict and abduction, and a food crisis that has left some 20 percent of the people in the area facing acute malnutrition.

How can we understand what has happened to the Lake Chad Basin (LCB) – or its relative invisibility in much of the wider world? (In late 2017 the *New Yorker* referred to the LCB as the site of the world's most complex, troubling humanitarian disaster,[1] yet outside the surrounding area and the confines of a handful of international aid organizations, few people know anything about it.) The situation is enormously complex. Long-term fluctuations in the size of Lake Chad are driven by natural forces, but its rapid shrinkage in the late twentieth century was also tied to the expansion of irrigated agriculture in response to population growth and a shift to larger-scale commercialized farming of export crops. Simultaneously, drought conditions have worsened in the face of a lethal combination of global climate change resulting from the burning of fossil fuels around the planet

3

and air pollution emanating from Europe that has affected air circulation patterns. Moreover, decades of poor governance and economic marginalization have made it difficult for many of the inhabitants to respond to changing conditions and have helped pave the way for the rise of the Boko Haram movement, which itself grew out of a more widespread turn toward radicalism in Southwest Asia and North Africa in the early 2000s. All of this has unfolded against the backdrop of a wider world in which most of those living in better-off regions pay little attention to the Sahel – a semi-arid zone of transition in Africa between the Sahara Desert and wetter regions to the south.

There is no easy way to unravel the complexities at play in the LCB, but it is impossible even to begin to grasp what has happened there without considering a few geographical fundamentals:

Location and _place characteristics_ matter. The developments described above are the product of a unique conjunction of environmental and human circumstances at a particular spot on Earth's surface. There is no other place on the planet where people are facing the same mix of human and physically driven environmental challenges: a decline in precipitation going back decades; the disruption of traditional flows of people and goods resulting from the creation of political boundaries and associated power dynamics; a lethal combination of violent insurgency and militaristic responses; and major socio-economic upheavals resulting from local ethnic divisions, the baggage of colonial arrangements, and the involvement of foreign governments and business concerns seeking to advance economic and political interests. The point is that the particularities of the geographical setting are of critical importance both to explaining what has happened there and to assessing the strengths and limitations of general understandings.

Human and physical processes are intertwined. The LCB is not facing either an environmental challenge or a human challenge; it is facing a combined human–environment challenge. This challenge manifests itself in a variety of ways. To cite just one example, the human and natural forces behind the shrinking of Lake Chad in the 1970s and 1980s created the perfect conditions for the expansion of the tsetse fly population, which led to an outbreak of disease in the cows on which island communities in the lake depended – in turn precipitating migration away from the islands that disrupted both the ecological and the ethnic balance in places where the migrants settled. If we only look at human or at physical factors, we can't understand what happened or why.

Spatial variations are revealing. We need to document and analyze the changing character of the LCB's physical and human landscape to assess what has happened and what might be done to address the crisis. By mapping and analyzing changes in the lake's surface, the surrounding vegetation, and settlement patterns using remote sensing and on-the-ground data collection, we can gain critical insights into the physical and human forces that are altering the character of the lake and disrupting the lives of the people who depend on it. Malnutrition is a much greater problem in some parts of the LCB than in others; looking at where the problem is more or less severe (i.e., its variations across space) can help us understand who is being affected – why, where, and how.

We need to look beyond the local. The LCB crisis is not simply a consequence of localized developments. Simplistic views of the crisis that focus solely on population growth, ethnic conflict, or resource management practices in the Basin fail to reckon with the myriad ways in which it is deeply affected

by developments originating far outside the region.
The human and physical forces behind the drought
are driven by global-scale processes. The colonial
order resulted in a political pattern that divided the
Basin into competing segments, which sparked, or
at least hardened, intra-regional animosities. The
expansion of water-intensive commercial agriculture was driven by consumption preferences and
economic arrangements emanating principally from
Europe. The rise of Boko Haram was inspired by
developments in Southwest Asia and found fertile
ground in an area long marginalized by outside
powers. In pursuing their geopolitical objectives,
France, the United States, and other powers have
helped solidify the power of corrupt state authorities; and they have responded to Boko Haram in
ways that have cost countless lives and threatened
economic stability in the region.

*Our understandings, priorities, and actions are
shaped by unexamined geographical assumptions.*
The limited attention paid to the LCB's challenges
by the outside world reflects a widespread tendency
in North America, Europe, and East Asia to relegate
Sub-Saharan Africa to the margins. It is difficult
to imagine that if a crisis of the same magnitude
occurred in southern Europe it would receive so

little attention. Paradoxically, even writing about the crisis as I have done here risks reinforcing the all-too-common, deeply troubling tendency among those outside of Africa to view the continent as a disaster zone, to ignore its enormous diversity, and to write it off as hopeless. The comparative global invisibility of what has happened in the LCB reveals the power of geographical imaginations to shape what gets attention, where resources are deployed, and how understandings develop.

The LCB crisis is, to be sure, extreme, but it is instructive of the types of circumstances that need to be taken into consideration when addressing developments in almost any place. It also serves as a signal example of the importance of bringing geographical perspectives to bear on issues and problems.⌈Geography is an academic discipline and subject of study that explores – and promotes critical thinking about – how the world is organized, the environments and patterns that exist on the ground or that humans create in their minds, the interconnections that exist between the physical and human environment, and the nature of places and regions.⌋ Geography, in short, offers a critically important window into the diverse nature and character of the planet that serves as humanity's home.

The Allure and Power of Geographical Understanding

Ever since early humans sketched primitive maps in the dirt, the quest for geographical understanding has helped people make sense of the world around them. Systematic assessments of the organization and character of Earth's surface allowed early scholars to figure out that the world was round; provided useful insights into where to locate settlements, plant crops, and find resources; promoted understanding of the workings of the physical environment; and, quite literally, helped human beings find their way. Through the ages advances in geographical understanding made it possible for people to explore the remotest corners of our planet and develop understandings of the interconnections that link the human and biophysical world. Like many bodies of knowledge, geography has served negative as well as positive ends: rapacious champions of colonialism used geographical knowledge to facilitate the exploitation of peoples and environments. Yet without some appreciation for geography we would be unable to comprehend how the world is organized and our place in it.

The search for geographical understandings traces its roots to human curiosity about other places. As the basic properties of Earth's surface became better known, attention shifted to what geographical arrangements could tell us about the planet – what, for example, the configuration of landmasses and the spatial arrangement of landforms reveal about the movement of tectonic plates; how the placement of political boundaries influences access to resources; how the organization of cities shapes people's activity patterns; and how the location of health clinics and grocery stores advantages some communities while disadvantaging others.

Since geographical arrangements are always evolving – cities expand, people move to new places, streams shift courses, the complex of economic activities in neighborhoods changes – the search for geographical understanding is never ending. Indeed, the importance of that search is growing because of the rate and extent of the geographical changes currently unfolding on Earth's surface. Sea levels are rising; large numbers of species are facing extinction; cities are exploding in size and population; the connections between distant places are being remade; people are moving around the planet at a previously unknown pace; record-breaking numbers of individuals are crowding into environmentally fragile places; and inequalities among and between places are escalating at an alarming rate. A recent study of the US National Research Council drew attention to the implications of these changes for geography:

> Stanford ecologist Hal Mooney has suggested that we are living in "the era of the geographer" – a time when the formal discipline of geography's long-standing concern with the changing spatial organization and material character of Earth's surface and with the reciprocal relationship between humans and the environment are becoming increasingly central to science and society.[2]

The critical importance of geography in the contemporary era also becomes clear when one considers the growing availability and use of maps and other types of geographical information to describe, classify, and analyze all sorts of phenomena. Geographical information systems (GIS) are now basic tools in everything from planning for emergencies to tracking migration flows. Global positioning systems (GPS) and computer maps have become a part of the daily lives of most people in better-off parts of the world. Accompanying and pushing these trends is a massive shift in the way many political and social institutions manage information. Until recently, most information about the world was organized by topic; now the dominant trend is to organize it by location or geospatial coordinate (the precise latitude and longitude where something is found).

Against this backdrop, it is not surprising that student interest in geography is growing, job opportunities are expanding, and a wide range of researchers and scholars are embracing geographical approaches and tools. To cite but a few examples, ecologists and biologists are using geographical techniques to map and analyze species distributions; social scientists are increasingly interested in the ways in which differences from place to place are fundamental to the social processes they study; a new transdisciplinary literature in law and geography is rapidly developing; hybrid disciplines such as geoarcheology and geolinguistics are attracting attention; and humanists are turning their attention to the importance of people's sense of place for the way they think about themselves and their relationship to others.

Despite these developments, the promise of enhanced geographical understanding is up against considerable public misunderstanding of the subject. Many people equate geography simply with memorizing place or location facts – where places are found and some of their defining features. Knowing these sorts of things has some value in that they make it possible to grasp basic features of Earth's surface and they allow people to situate themselves in relation to other places and peoples. Yet if the case for geography were to rest primarily on knowing selected geographical facts, it would be a rather flimsy case indeed – particularly in an age when 30 seconds on the internet can produce an answer to most basic locational and place-fact questions.

[Geography, however, is much more than that. At its heart, modern geography is concerned with studying the arrangement and character of Earth's surface – the spatial organization of phenomena found there, the intertwined physical and human systems that shape its features, and the nature and meaning of its constituent places and regions.[3]) Figure 1 offers a useful window into geography's fundamental orientation. The diagram points to geography's concern with environmental, societal, and coupled human–environment systems (the vertical axis of the cube); its emphasis on what can be learned about those systems from the study of places, patterns, and scale (one of the bottom axes of the cube); and its use of a variety of types of spatial representations to advance understanding (the other base of the cube).

Geography's perspectives and tools shed light on

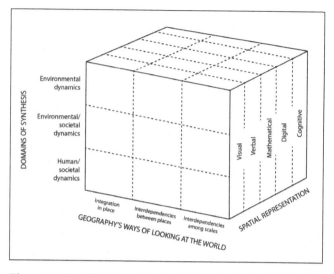

Figure 1 Visualization of geography's substantive foci, perspectives, and approaches.

where things happen, why they happen where they do, and how geographical settings influence physical and human processes. As part of that effort, geographers pay attention to the spatial organization of Earth's surface – exploring what can be learned from studying locations, patterns, and distributions (drought patterns, disease clusters, ethnic distributions, etc.). They consider how different processes come together to create distinctive spaces, places, and regions – creating a world in which no two places are exactly alike. And they investigate how those distinctive spaces, places, and regions produce particular geographical contexts – physical and social, material and imagined – that reflect and shape developments on the surface of the planet. Whatever the point of entry (there is much overlap among the foregoing areas of emphasis), geography is centrally concerned with what is sometimes called the "why of where"[4] – the nature and significance of differences across space; the impacts of place-based circumstances on environmental, social, and human–environment systems; and the influence of *where* something happens on *what* happens.

Given this intellectual orientation, it is no surprise that the objects of geographical inquiry encompass everything from the dynamics of glaciers to migration patterns, from the diffusion of pests in forests to the relationship between ethnic and political patterns, and from people's sense of place to the forces promoting segregation in cities (e.g., social and class prejudices, the practices of lending institutions and real estate agents and gentrification pressures). It follows that geography is best understood as a discipline united more by a set of shared perspectives than a particular topic of study – an orientation that makes it more akin to history than to most other disciplines. Historians work on everything from the expansion of the ancient Persian Empire to the social impacts of Germany's mid-twentieth-century "guest worker" program. What binds historians together is not a topic but an interest in understanding the evolution of the human story and its implications for the present. The equivalent for geography is an interest in the patterns, environments, and places that make up our planet – what they are like, how we think about them, and how they influence people and nature.

Geography's Historical Roots

To appreciate the character of geography, it is useful to see how the body of ideas and perspectives associated with the discipline developed over time. All societies have sought insights into their surroundings in ways that resemble what we now commonly term geography, but the term itself, and its institutionalized practice throughout most of the world today, is a product of a formal branch of knowledge that can be traced back to the ancient Greeks (_geo_ = from the Greek word for Earth; _graphy_ = from the Greek verb to write). As Greek explorers left their homeland more than 2,000 years ago and began traveling along the shores of the Aegean and Ionian seas, and then moving on to more distant lands, they encountered rivers that were larger and more powerful than any they had seen before, trees they had never encountered, and peoples speaking languages very different from their own. They made maps showing the places where they had been and wrote accounts of what they found. They were not just interested in cataloging, however; they also asked geographical questions. Why are certain plants found in one place but not in another? Why do some rivers flood every year and others do not? Why are customs different from place to place? Which locations are strategically advantageous for the development of cities? Addressing those questions allowed them to develop fundamental insights into the diffusion of biophysical and cultural traits, the relationship between precipitation and stream behavior, the nature of Earth as a planetary body, and the advantages and disadvantages of different locations for settlements.

Geographical knowledge did not serve solely benign ends; it also facilitated conquest and the construction of an expansive empire. What is clear, though, is that for the Greeks geography was not just a set of place facts; it was a subject that helped them make sense of their place on the planet through the careful cataloging *and analysis* of the patterns (both human and physical), places, and environmental contexts they encountered. That sense of geography endured through the ages. The ancient Romans drew on Greek ideas to organize information about (and rule) their expanding empire. Persian and Arab geographers picked up the geographical mantle in the West after the fall of Rome – refining calculations of latitude and longitude and advancing understanding of the organization and character of the known world at the time. The formal study of geography then found its way back into the European heartland in the late Middle Ages and early modern period, helping to give rise to the Enlightenment.

Geography's development in the West was paralleled by the emergence of increasingly sophisticated geographical understandings in other parts of the world – most obviously in China. The European colonial project, however, ensured that the Western tradition became the most broadly influential one. To be sure, the practice of geography was often dominated by mechanical concerns; the same was true of most fields of inquiry. The Romans are particularly noted for their precise mapping of new lands and their atlases, the Islamic geographers of the Middle Ages for their sophisticated cartographic productions and their descriptions of newly discovered lands, and the European geographers of

the fifteenth to seventeenth centuries for their com-
pilations of the attributes of places that were the
targets of colonial ambition. Indeed, the importance
of factual geographical knowledge for the creation
and maintenance of colonial empires is undeniable.

Yet across the centuries there were always individuals who were not just amassing geographical
information; they were asking insightful geographical questions and using geographical information
in the service of understanding the world around
them. One of the best-known Roman geographers,
Claudius Ptolemy (AD 100–70), did not simply assem-
ble geographical facts; he advanced ideas about the
nature and consequences of climate patterns and he
developed a grid-system approach to mapping that
is still with us today. The eleventh-century Persian-
Muslim scholar Abu Rayhan Al-Biruni proposed
innovative ideas about the potential for different
natural zones on Earth's surface to sustain human
life. Studies of the configuration of the coastlines on
either side of the Atlantic led the sixteenth-century
Flemish geographer Abraham Ortelius to propose
that the continents might have drifted apart – an
idea about "continental drift" that only came to be
widely accepted in the latter part of the twentieth
century.

Given this history, it is not surprising that, with

the rise of modern Western educational systems, geography became a core component of the primary and secondary school curriculum, and also found its way into universities reorganizing themselves along disciplinary lines. These developments led to the emergence of academic geographers who began undertaking systematic studies of the workings of the physical environment, the interrelationships between humans and the environment, changes in the distribution of peoples, the organization of settlements, the spatial arrangement of economic activity, and patterns of inequality.

Geography's place in the modern university owes much to the influence and prestige of three eighteenth- to nineteenth-century Germans. The first was Alexander von Humboldt (1769–1859), who conducted detailed analyses of the geographical character of plant communities in Latin America that provided the first real understanding of the influence of the physical environment on vegetation and gave rise to the field of biogeography. The second was Carl Ritter (1779–1859), who wrote a nineteen-volume treatise discussing the influences of the physical environment on human activities in different world regions and looked at the geographical organization of human practices and institutions, including the state. The third was Immanuel Kant

(1724–1804), who championed the teaching of geography as fundamental to human betterment and saw the discipline as a vehicle for integrating knowledge about the world.[Kant treated geography as a subject concerned with "the difference that space makes" in mathematical, moral, political, commercial, and theological realms.[5]]

Kant had an enormous impact on the modern university. His ideas about the nature of knowledge and human perception provided the intellectual basis for the disciplinary matrix that emerged in nineteenth-century German universities. Those universities, in turn, greatly influenced institutions of higher learning elsewhere, especially in the United States. In his *Critique of Pure Reason* (1781), as well as in other writings, Kant made a case for studying three things the human senses could grasp: phenomena that can be grouped together because of their objective qualities (i.e., different "forms" or topics), time, and space.[6] The newly organized disciplines taking root in nineteenth-century German universities reflected this way of thinking: a series of topically defined disciplines such as botany and politics (form), history (time), and geography (space). Later critics argued that separating time and space was intellectually indefensible, but Kant was not championing a discipline of geography focused solely on spatial matters; he saw time and space as mutually reinforcing, meaning that geography and history should be thought of as two sides of the same coin.

By the early twentieth century, geography had a foothold in many institutions of higher education in the West: in Germany, France, the United Kingdom, Russia, the United States. By the second half of the century, geography departments reflecting the Western geographical tradition had sprung up at institutions of higher learning all around the world, and geography came to be viewed widely as a core discipline. Its mid- to late-twentieth-century institutional status in the United States, however, was more checkered. Geography teaching in schools was in decline as it came to be folded into a more general "social studies" curriculum, and geography itself was mired in a descriptive phase. Several prominent American universities abandoned geography – most notably Harvard in the late 1940s – and many of the country's liberal arts colleges, along with some public universities, came to view it as a dispensable part of the curriculum. Such developments, in turn, fostered a misguided perception of geography as an old-fashioned subject – a perception further fueled by the male-dominated character of the discipline at the time.

This way of thinking is increasingly being challenged (references to geography's rediscovery are not uncommon[7]), but breaking out of the older mindset is still a work in progress. Even in Europe – Western geography's intellectual homeland – shifting priorities and a proliferation of new types of interdisciplinary initiatives have challenged geography's traditional status and standing. Moreover, a popular view persists equating geography with place-name memorization. That way of thinking stands in stark tension, however, with the perception of students on campuses with strong geography programs who encounter a discipline giving them insights into the interconnections affecting the fates of places around the globe; expanding their understanding of environmental processes and human–environment interactions; deepening their appreciation of how those processes play out differently from place to place; and offering them the skills with which to use, and evaluate the advantages and limitations of, geospatial technologies.

Geographical Thinking

Geography is such an exciting, relevant realm of inquiry because of the insights that follow from looking at issues and problems through a geographical lens. Consider the case of the Soviet invasion of Afghanistan in 1979–80. The invasion came as a surprise to most commentators, who scrambled to come up with reasons for it. As one of my former colleagues liked to recount,[8] soon after the event media and policy discussions began to focus on two possible motives for the Soviet Union's actions: (1) the invasion was a first step toward securing a long-sought-after Russian warm-water port on the Indian Ocean (since Afghanistan is landlocked, the implicit suggestion is that the invasion of the country would be followed by a move to the Indian Ocean through sparsely populated southwestern Pakistan); or (2) the invasion was in keeping with a long-standing pattern of Russian territorial expansion, so it should be seen as a step on the way to the incorporation of Afghanistan into the Soviet Union as an additional Soviet Republic. Each of these explanations was given extensive coverage in the media at the time, and they both received serious attention in policy circles.

For casual consumers of the news, these explanations might have had a superficial appeal. Yet applying even basic geographical reasoning exposes fundamental flaws in each of them. A quick look at a map of Pakistan shows no major ports along the southwest coast. There is a reason for that: the continental shelf there is so shallow that large ships cannot easily gain access to the coast (hardly a prize worth fighting for given the major international conflict that would be required). As for incorporating Afghanistan into the Soviet Union as an additional Soviet Republic, a glance at an ethnic map and a nodding acquaintance with the growing restiveness of the Soviet Union's Islamic population would have revealed that this would involve the incorporation of 18 million additional Muslims into the Soviet Union, which was one of the last things the authorities wanted. (As subsequent events made clear, what the Soviets were really trying to do – unsuccessfully, of course – was to create stability on their southern flank by exerting the kind of control over Afghanistan that they had over much of Eastern Europe.)

What this example shows is that geographical thinking – together with awareness of some basic geographical facts – can offer illuminating insights. Rather than focusing solely on institutions, powerful figures, or ideologies, looking at an issue through a geographical lens draws attention to the role played by underlying spatial patterns, environmental circumstances, and locational characteristics.

Geographical thinking also opens up questions

about the nature and soundness of the geographical representations people use to describe events and developments. It is almost impossible to talk about any development taking place on our planet without framing it geographically. A discussion of environmental issues in "the Middle East" or "Europe" draws on large-scale geographical constructs – in both cases ambiguous ones in terms of their territorial extent. A ranking of countries based on per capita GNP invites people to organize economic information on the basis of units shown on the typical world political map rather than other types of spaces that might be more comparable (think of what it means to treat Russia and Luxembourg as comparable units). A magazine article comparing the health-care picture in Cornwall and Essex encourages thinking about differences at a particular spatial scale – the scale of England's counties – in the process telling a different story from the one that would be told if a different scale of analysis were used. A map in a newspaper showing clear-cutting in US National Forests invites consideration of what is happening in an area with distinct boundaries while directing attention away from relevant cross-boundary circumstances with significant ecological impacts. In the absence of geographical thinking, it is easy to overlook what is hidden and what is revealed when an issue is framed against the backdrop of a particular geographical space or scale. Studying geography, in other words, is a horizon-broadening experience.

Studying Geography

Academic students of geography learn to understand and think about the world around them by exploring climate, politics, biology, economics, and other topics through a geographical lens. They also learn about various techniques and tools that are integral to geographical inquiry. In the modern academy, the search for geographical understanding is rooted in diverse, sometimes competing, theoretical orientations and perspectives (positivist, humanist, Marxist, post-structuralist, feminist, etc.). Each offers different insights into geographical arrangements and processes, and at times they point to sharply divergent interpretations. Yet at times each also has informed the others in ways that have contributed to the advancement of geographical understanding.

Geographical inquiry also draws on a diversity of methods: collecting and analyzing information about locations, patterns, and distributions;

documenting and evaluating changes in the land-scape (the geographer's approach to the landscape is loosely analogous to the way literary scholars approach texts – thinking critically about what tangible forms reveal about underlying processes); and looking for evidence of the forces that shape the geographical character of the planet. Not sur-prisingly, for many geographers maps, images of Earth's surface (air photos, satellite data, etc.), and computer-based GIS are important geographical tools because they can provide insight into spatial patterns and facilitate decision-making. Figuring out the best route for a path traversing a wildlife habitat area is no easy task because a variety of factors must be taken into consideration: wildlife migration patterns, terrain characteristics, the distribution of plants, pre-existing pathways, and vulnerability to natural hazards. Each of these has spatial characteristics that can be captured using GIS and assessed collectively to determine which routes make the most sense. GIS are also used to create visualizations that provide evocative insights (see, e.g., plates 4 and 9). The explosion in new geographical technologies has been so great that a new field of inquiry – geographical information science (GISci) – has come into being. Practitioners of GISci focus not just on GIS applications but on the characteristics, potentials, and limitations of geographical technologies.

For all the importance of cartography and geo-spatial technologies, geography is not just about making or interpreting maps. Indeed, many geo-graphical studies make little use of maps. Careful observation, photographs, and field notes provide information about landscapes; ethnographic meth-ods, interviews, and surveys offer insights into the characteristics of places, the people who live there, and the significance of differences from place to place; the textual analysis of media reports and personal communications sheds light on how issues are geographically framed; and philosophical reflec-tions about the nature and meaning of place and space can open up new ways of thinking about ourselves and our roles in the world.

The study of geography is wide-ranging, but that does not make it incoherent. Its concerns and meth-odological approaches revolve around a meaningful constellation of understandings, ideas, techniques, and approaches. To gain a deeper appreciation of their nature, and to explore what they have to offer, each of the next three chapters delves into what it means to adopt geography's core objects of analyti-cal attention: spatial patterns and arrangements, the character of places, and physical–human interactions. There is a great deal of overlap among and between these objects of study, but examining them individually provides a useful means of demonstrating what it means to look at the world through a geographical lens. Following these chapters, attention turns to the importance of the discipline's general educational mission. The book concludes with brief consideration of the importance of promoting greater understanding and appreciation of modern geography in our fast-changing world.

2

Spaces

One word is inextricably associated with geography: where. That is because geography starts from the premise that it matters where something takes place on Earth's surface. The key questions are not simply "where" questions, though; they are "why there" and "so what" questions. Getting to such questions means taking spatial arrangements, variations, and interconnections seriously. Engaging in even the simplest day-to-day activity requires some appreciation of spatial circumstances – where to find food and services, how to get to work places, and the like. Moving up in scale, without some awareness of how phenomena are arranged on Earth's surface, it is difficult to make reasoned business or policy judgments, make sense of events, or grasp some of the basic forces shaping life on the planet. Locating a new store or public service requires taking into consideration population distributions, the location of roads and utilities, socio-economic patterns, and more. Understanding why and where migration happens requires consideration of the political organization of territory, the spatial consequences of discrimination, socio-economic patterns, and the layout of the physical environment.

Studying spatial arrangements is sometimes the best, or even the only, way to gain insight into vexing scientific and human problems. No one was quite sure what caused cholera epidemics until the British physician John Snow made a map identifying reported cases of the affliction during an outbreak in London in the early 1850s. The map showed most of the cases clustering around a single well – concrete evidence in support of his idea that cholera was a water-borne disease. Analyzing the geographical distribution of diseases, as well as changes in those distributions over time, is now a fundamental tool in epidemiology.

Spatial analysis is equally important in other arenas. Figuring out the human contribution to climate change is extraordinarily challenging because of the large number of factors affecting climate and their complex interactions and variability. Because natural forces have driven climate change throughout Earth's history, the only way to determine the extent of humanity's recent contribution is t develop models of how the climate system works over the long term, and then see how much present conditions deviate from what the models predict. Spatial analysis has played a critical role in that endeavor.

Some of the major natural drivers of climate change are well known – variations in the eccentricity of Earth's orbit, shifts in the degree of tilt of its axis, and the procession of the equinoxes (all three of these control the amount of incoming solar radiation). Other influences include variations in solar activity, the composition of the atmosphere, and alterations in the reflectivity of Earth's surface and atmosphere. Climate models seek to capture how the mix of these ingredients plays out over time. Detailed reconstructions of past vegetation patterns have proven to be of great help in the creation and refinement of climate models because vegetation patterns reflect climatic circumstances. Hence, determining what those patterns were like in the past, and how they have changed over time, can yield important insights into the workings of the climate system.

One way to reconstruct past vegetation patterns is to take sediment cores from the beds of lakes and examine the types of pollen found in the various layers of the core. (The oldest layers are at the bottom and the youngest at the top, and the layers can be dated using a variety of modern dating techniques.) The types of pollen found at a particular level in a sediment core provide evidence of the kinds of vegetation (and by extension the climate) that existed in the surrounding area at the corresponding time period. Scientists with a geographical bent have used evidence of this sort from multiple sites to construct maps of changing vegetation and climate in different regions during the Quaternary (the past c. 2.5 million years). Comparisons of maps such as the top one in plate 2 – a climate reconstruction map – with maps depicting climate-model predictions (the bottom map) make it possible to test, and repeatedly improve, the models. The top map, for example, shows the reconstructed precipitation of 6,000 years ago based on pollen data, with blue indicating locations that were wetter than today and brown locations that were drier. The bottom map shows the simulated differences in precipitation between 6,000 years ago and the present, colored the same way. Comparison of the maps shows that the major differences in reconstructed precipitation between then and now are also simulated by the climate models. However, there are some differences that we can learn from in future work. The simulated precipitation in the center of Asia is a little too low 6,000 years ago. This provides insight into adjustments that can improve the models.

The increasingly good job climate models do in simulating the past raises confidence in their accuracy and usefulness. Moreover, the fact that climate models are unable to simulate the climate of the last few decades accurately without factoring in human influence provides strong evidence that humans are significant agents of contemporary climate change – and a powerful response to climate skeptics who contend that climate change is simply a natural phenomenon over which humans have little influence. This example shows that the geographical concern with spatial analysis extends far beyond simply locating phenomena on Earth's surface; it is a fundamental tool for analyzing and addressing matters of profound contemporary importance. Studies of the changing distribution of tree species in forests shed light on the degree to which climate change is altering ecosystems in different regions; the detailed mapping of population distributions in coastal regions reveals which areas are most vulnerable to earthquakes and tsunamis; and analyses of ethnic distributions in conflict-torn places increase awareness of the threat of violence faced by different communities. Against this backdrop, it is easy to understand why the use of GIS has expanded rapidly, for they facilitate efforts to understand the interrelationships among different factors on the ground.

Geographers approach the study of spatial arrangements in several basic ways. They seek to identify and explain the significance of spatial patterns. They explore what variations across space tell us about the forces shaping biophysical and human processes. They investigate the nature and meaning of interconnections across space and scale. And they look critically at the spatial ideas and frameworks humans use to understand, navigate, and seek to change the world around them – asking what those ideas and frameworks reflect, how they influence understandings and practices, and how they might be altered in the service of human betterment and environmental sustainability.

Spatial Patterns

How severe is the bark beetle infestation in Western North America, and how might it be curtailed? What might be the cause of higher-than-normal reported cases of cancer in a particular neighborhood in Dublin? What obstacles do poor immigrant communities around the city of Paris face in accessing food and public services? A useful first step to answering any of these questions is to figure out the spatial character of the phenomenon in question – where forest tree death is particularly acute, where people with cancer diagnoses are clustered, where food stores and clinics are located in relation to population concentrations. An interest in answering "where" questions has given the map a special place in the work of many geographers, for mapping is an effective way to make sense of spatial patterns and promote an understanding of them.

Some maps tell powerful stories at face value, but the importance of making a map often lies in the questions it raises and the routes to finding answers it suggests. A map showing the distribution of different ethnic groups within cities can offer insights into why some groups are more segregated than others, but such a map can also suggest the types of factors that might be responsible for patterns of segregation. Mapping streambank characteristics along a river can paint an interesting picture of wetland habitats, but it can also lead to hypotheses about the causes of erosion at particular places along a stream's course.

Of course maps are not value-free depictions of reality. They reflect particular perspectives and priorities (a point we will explore in more detail later), and they therefore are appropriate targets for critical examination. Consider, for example, plate 3, a map that juxtaposes a proposed early-1990s partition plan for war-torn Bosnia and Herzegovina (the regions shown in colors) with a set of regions depicting how people moved around and used space prior to the outbreak of the conflict (the solid and dashed lines). This map was created to make a point – to show that the partition plan proposed by the United Kingdom's Foreign Secretary, Lord Owen, and his US counterpart, Cyrus Vance, represented a problematic approach to dealing with the conflict that pitted Serbs, Croats, and Bosnian Muslims against one another. Their plan did not ignore spatial variables; it was based on a map dividing the country along administrative district lines, with districts allocated to one or another group depending on which group had a plurality in each district. In an effort to show why those were not good spatial variables to use, Austrian Academy of Sciences geographer Peter Jordan used information about commuting patterns in pre-conflict Bosnia to show that the spaces that were meaningful in people's lives – what he called micro and macro "functional regions" – did not correspond to the district map of ethnic pluralities.[1] Plate 3 drives the point home and is suggestive of why the plan ended up failing (it was soundly rejected by all sides). Even this map does not tell the whole story. Bosnian Muslims would have ended up with far less productive agricultural land than their Serb or Croat counterparts if the plan had gone into effect – another insight that comes from looking at the plan through a geographical lens.

Mapping and other representations of spatial data, then, should generally not be thought of as ends unto themselves; they are the product of efforts to understand what gives rise to particular patterns and to explore how those patterns influence what happens. Such efforts are always partial and open to question (the questions that are asked and the data that are examined influence what is found), but they can be extremely valuable. They can shed light on why some glaciers are advancing while most are retreating, what places are more and less vulnerable to natural hazards, how the layout of a transportation grid advantages some communities while disadvantaging others, and why those living in certain neighborhoods face greater social, economic, and environmental challenges than those living elsewhere.

Spaces

Variations across Space

Another way of thinking about the difference that space makes is to consider what can be learned by looking at variations from place to place. In the search for general laws governing physical and human systems, scientific studies sometimes treat those differences as little more than noise obscuring more general processes. Many economists and political scientists, for example, assume in their models that people everywhere respond to developments in similar ways (making "rational choices" or acting in their "self-interest" – both treated more or less as human universals). And biologists and physicists sometimes assume that the workings of the physical environment can be most effectively revealed if they look for similarities across many places rather than focusing on localized conditions.

There are certainly insights to be gained from studies that apply general laws to explain phenomena. This is particularly true in the physical-environmental realm, where understandings of everything from gravity to pressure differences in the atmosphere are products of the search for such laws. In the human arena, the analysis of financial flows or the spread of diseases through human contact can also benefit from that kind of analysis.

But taking a geographical approach also means looking at what can be learned by foregrounding differences across space – considering, for example, how geographically specific circumstances influence stream-bank erosion or decisions to migrate in the face of economic hardship. To the extent that broader generalizations can be made, they come from comparing individual cases and teasing out what is general and what is specific.

Focusing on variations across space can draw attention to the limitations of sweeping generalizations such as the one that forms the premise of *New York Times* columnist Thomas Friedman's influential 2005 book, *The World is Flat*.[2] In that book Friedman argues that globalization is creating an increasingly geographically undifferentiated planet because high-tech workers in Silicon Valley, for example, are now inextricably tied to, and in competition with, their counterparts in London, Amsterdam, Tokyo, Bangalore, and beyond. Friedman sees other reflections of his flat world in business elites from around the world traveling and living far from their homelands, professors from multiple countries communicating regularly and collaborating on projects, the wealthy from all over the world vacationing in similar places, and British Premier League football teams being followed with great intensity in southern Thailand as well as northern England.

Globalization has certainly brought people together in unprecedented ways, and for certain people in certain places, some leveling of the playing field has arguably taken place. Nonetheless, thinking geographically about Friedman's thesis quickly exposes its limitations. By some estimates close to half of the world's people will never travel farther than 100 kilometers (c. 60 miles) from the place where they were born. The gap between rich and poor has widened in many places. The life of a rural dweller in northern Yemen is about as far from that of his or her counterpart in rural Scotland as could be imagined. Even differences across modest distances divided by the boundaries separating countries can be stark. The opportunities and prospects defy comparison for a child born into a middle-class family in Paju, South Korea, and one born in Kaesŏng, North Korea – a mere 25 kilometers (15 miles) away.

One of geography's most valuable contributions lies in the attention it draws to these differences. Plate 4 provides a powerful counterpoint to the world-is-flat narrative. That map uses different colors to show travel time to population centers: areas shown in yellow have the greatest access to the types of developments highlighted by Friedman, whereas those shown in dark red and purple have the least. Since access to urban areas is strongly correlated with health and socio-economic indicators, particularly in low- to middle-income areas, the map tells a story about the variable opportunities and challenges people face in the world today that is largely absent from Friedman's account (indeed it is hidden from view in conventional maps depicting levels of economic development based on country-level GDP per capita figures).

When the types of differences depicted in plate 4 are considered alongside the failure of many generalized models to adequately explain or predict matters ranging from economic downturns to the intensity of floods, the importance of geography's interest in variations across space comes into focus. The case of Siberia's shrinking lakes provides a case in point. There is clear evidence that Siberia's climate is warming, resulting in permafrost melt. It was long assumed that melting permafrost produces an expansion in the number and size of lakes – particularly in a region such as Siberia where precipitation has marginally increased. Yet when UCLA geographer Laurence Smith and colleagues used remote sensing data to examine in detail the changing spatial arrangement and physical character of Siberian lakes, they found something different: a significant decline in the number and size of lakes in parts of the study area – specifically lakes lying near the permanent permafrost boundary.[3] That insight led them to posit that lakes in that location do not conform to general assumptions; instead, permafrost melting leads initially to lake expansion, but then those lakes tend to drain as the underlying complex of sediments, soils, and rocks becomes more permeable. Studies in this vein will surely be critical to understanding what areas will be more or less affected by climate change in the decades to come.

Turning to a different type of example, creating safer, more livable cities is a widely championed goal, but policies that promote this end in some places may not have the same effect elsewhere because local circumstances differ. In the aftermath of World War II, local governments in southern Louisiana began adopting national building code standards that aimed to promote the development of more durable built environments across the country. One of the standards called for slab-on-grade construction (concrete slabs laid down on the ground that serve as the foundation for structures, with no open space between the ground and the structure). Following this standard in southern

Louisiana meant abandoning the older practice of building houses on piers (traditionally these were 18–24 inches above the ground). During the great floods that hit Baton Rouge in August 2016, a considerable majority of the houses that were damaged were newer slab-on-grade constructions that were inundated by less than 18 inches of water. In other words, the failure to consider the particularities of individual places – in this case the high risk of flooding in southern Louisiana – led to grave consequences for tens of thousands of people.

The only way to avoid consequences of this sort is to take variations across space seriously. A public transportation plan that works well for a relatively compact UK city such as Southampton is not necessarily appropriate for a more sprawling Manchester. Active fire suppression policies that make sense in vulnerable parts of the Australian bush are not necessarily appropriate in remote parts of Canada's boreal forest. The point is that minimizing the significance of local environmental, demographic, social, and cultural circumstances – looking past them in the effort to develop more general understandings or widely applicable policies – can easily impede understanding and work against the development of effective responses to social and environmental challenges.

Spaces

Interconnections across Space and Scale

Places do not exist in isolation. They are affected by circumstances and events near and far, and those effects have grown as the world has become ever more interconnected. The strong yes vote in favor of Britain leaving the European Union (EU) in the northeastern coastal English town of Hull was not simply the result of local circumstances. Instead it reflected a sense of economic marginalization relative to other places, a feeling that ruling elites in London had long ignored the hardships of northern England, and frustration over a decline in the fishing industry that many attributed to the EU.[4] It also was the product of local unease over immigration from Eastern Europe and a politically driven effort at the national scale to cast the United Kingdom as a country that no longer controlled its own affairs and that had fallen victim to unchecked immigration.

The Hull example shows why connections across space are so critical to making sense of what happens in a given place. The issue is not simply the interconnections that exist among discrete locations. Connections across scale matter as well. In the Brexit case the national and European scales affected the local – and of course what happened

46

at the local scale had broader implications as well (Brexit was the product of multiple places such as Hull voting in a particular way). Looking at interconnections across scales – spaces of different size ranging from the local to the global – helps to put places, events, and processes into context and reveals important causal connections.

The importance of spatial interconnections comes into focus when one stops to consider that even attributes typically associated with individual places are often the product of interconnections across space and time. When asked to think about Switzerland, many people immediately think of chocolate, yet cocoa beans have never been grown there. Of course dairying has long been important to the country's agricultural economy, but it was Switzerland's ability to develop and control trade in a commodity produced far from its borders that gave it one of its signature products.

Over the past century the economies of all but the most remote areas have become increasingly inter-twined with other places. Many a retail complex specializing in clothing in suburban Washington, DC, sells products that are part of an intricate "commodity chain" characterized by, for example, the production of cotton in Central Asia, which is then shipped to Turkey for weaving into yarn, which is then taken to China for coloring, which is then moved to Vietnam, where it is combined with buttons from France and sewn into a pair of trousers, which is then shipped to the Washington suburbs for sale – often with a label declaring it was designed in the USA or a place known for fashion such as France or Italy. Various forms of foreign ownership and control are present at different links along such commodity chains, and they are implicated in the transportation of goods that make these chains possible. As such, most commodity chains are part of even more complex "global production networks."[5]

The importance of paying attention to such networks came into focus in the lead-up to the January 2017 inauguration of US President Donald Trump. Trump suggested that his government should consider imposing a 35 percent tariff on automobile imports from Mexico as a means of protecting US jobs. Whatever one might think of the wisdom of protectionist policies more generally, this policy proposal was seemingly advanced without any consideration of the interpenetration of the US and Mexican automobile industries. In fact, more than a third of the parts incorporated into US-assembled cars come from Mexico, and cars assembled in Mexico contain many parts made in the United

States – meaning that, if implemented, the proposed tariff would have a strongly negative impact on automobile production, and by extension jobs, in both countries. Simply put, any reasoned discussion of tariff proposals and their potential impacts cannot take place without due consideration of relevant geographical connections across space.

The economic arena is not the only one where connections across space and scale are important. The vegetation complex along the shores of Lake Como in Italy reflects a mix of local, regional, and larger-scale biophysical circumstances, the importation of exotic species by humans, and the impacts of agricultural practices, human-caused pollution, and the expansion of settlement around the lake. The conflict that has wracked Syria in the second decade of the twenty-first century is the product of external colonial practices (boundary drawing, externally imposed administrative structures, the creation of an export-driven economy, etc.), complex interactions with neighboring countries (especially Turkey and Iraq), Syria's niche in regional and global production networks, and the geopolitical ambitions of Iran, Russia, the United States, and other powers. The conditions of a shantytown in Durban, South Africa, are rooted in colonial-era power relations, the inertia of a long-standing racist regime presided over by a ruling elite in Pretoria, regional and global economic structures that have enriched some sectors of society at the expense of others, and so on.

Interconnections across space and scale are so dizzyingly complex that, in most instances, it is impossible to tease out the full range of forces, near and far, that produce particular outcomes. There is, moreover, much debate among geographers about what deserves attention and what theories are best equipped to explain geographical outcomes. The effort to identify and describe the connections shaping geographical circumstances, however, is a prerequisite to any productive consideration of, or debate about, those connections.

Given the increasing intensity of connections shaping the planet in the twenty-first century, geography's long-standing concern with understanding the ways in which developments unfolding at one place or scale affect what happens at another could not be more important. What are the implications of consumption patterns in Europe and North America for farming practices in Africa and South America? How are land-use practices in the upper reaches of the Yangtze watershed influencing flooding downstream? How are new virtual interaction technologies changing patterns of movement and altering the organization of cities? To what extent are commercial fishing practices, the construction of new pipelines, and oil fracking altering the fates of indigenous communities? Addressing geographical questions such as these will be critical to confronting many of the issues of our time. They also have the potential to promote critical thinking about the places and scales that are used to frame discussions of particular issues. If, for example, drug-related violence in Mexico is thought about not just as a Mexican issue but against the backdrop of the larger geographical reach of the drug trade, it would be harder to ignore the impacts of drug use in the United States and Europe on social stability in Mexico.

Questioning Spatial Assumptions

The Mexico drug violence example points to the importance of critical thinking about spaces and spatial arrangements. Almost every discipline or area of inquiry seeks to bring a critical perspective to bear on its objects of study. For geography that means not simply taking spatial ideas for granted, but also questioning their value, utility, and appropriateness. It means being aware that an analysis focused on socio-economic issues at the scale of cities or neighborhoods as opposed to states or countries invites different understandings of causes and consequences – even if one is attuned to the influence of larger-scale processes on smaller-scale places. It means recognizing that an assessment of the environmental challenges facing the Mediterranean region will draw attention to different things than will an analysis focused on either southern Europe or northern Africa.

Journalists, politicians, policy-makers, educators, business leaders – indeed all people – invoke geographical constructs as they describe, and seek to act upon, the world. Sometimes the parameters of these constructs are vague (the Middle East, the American Midwest) and sometimes they are more precise (Australia, metropolitan London). Whichever is the case, it is always worth considering whether the construct makes sense as a frame of reference – and reflecting on what is hidden and what is revealed when deploying a given framework. It is equally important to recognize the degree to which the availability of certain kinds of information or data shapes what gets attention and what does not. Since data often tend to be collected and aggregated along political jurisdiction lines, for example, far more studies focus on developments within units on either side of political boundaries than on those that straddle them.

The political pattern's influence on data gathering and dissemination helps explain the general tendency to treat the map of independent countries simply as a set of pre-given spaces that are beyond question. Few commentators even stop to consider how much the political map dominates the modern geographical imagination. For every reference made to what is going on in the Congo River Basin, the French-language region of Europe, or the wheat belt of North America, thousands and thousands are made to the Democratic Republic of the Congo, France, and Canada. It is a commonality to describe the location of developments – even those that are not political – by referencing one of the units on the world political map (the tsunami happened in Japan rather than the northeast coast of Honshu), to describe people in terms of the countries from which they hail (she's Bolivian, not Quechua, Guaraní, or Aymora), and to organize most information about the world along state-territorial lines (the literacy rate in India is 30 percent lower than in China).[6]

In the absence of critical geographical thinking about the pattern depicted on political maps, it is easy to ignore important disconnects that exist between the political pattern and other geographical patterns (demographic, ethnic, environmental, etc.) or to treat all states as if they were essentially the same thing (e.g., viewing China, with close to 1.4 billion people, a land area of more than 9 million square kilometers [c. 3.5 million square miles], and a massive bureaucracy, as the same kind of entity as the tiny South Pacific island state of Nauru, with a population of less than 10,000, a land area of around 21 square kilometers [just over 8 square miles], and a state apparatus that is dwarfed by the municipal government of just one modest-sized Chinese city). To be sure, events such as the break-up of the Soviet Union, the disintegration of the former Yugoslavia, or the success of separatist movements in places such as South Sudan serve as periodic reminders that the political pattern is not static. But without critical thinking about the pattern shown on the world political map, such events are unlikely to challenge the idea that the map sets forth an unproblematic framework for locating and describing events and processes.

In contrast, developing even a modest habit of thinking geographically can bring a host of revealing questions to the fore. Why do we typically show Somalia as one country on world political maps when the north and the southeast function entirely independently of one another? What are the consequences of Nepal being sandwiched between India and China, with much easier access to the former than the latter? Questions such as these – geographical questions – can foster the type of thinking that is essential to making sense of the contemporary geopolitical scene.

Critical thinking about space is also important to addressing smaller-scale tangible and applied problems. Are the boundaries of an area set aside to promote species conservation in the right place? Are the streets and sidewalks in a downtown area configured to promote accessibility and ease of movement? Does the location of bus stops, train stations, and highway access points favor some groups over others? Are decisions about the location and internal organization of public parks sensitive to the interests and constraints of different segments of the surrounding population? No urban form can ever be completely equitable, but asking these and related questions is crucial if we are to create more livable, just, sustainable communities in the years and decades ahead.

Another set of valuable insights comes from thinking critically about the ideas underpinning the ways in which particular maps are drawn, and the role those maps play in shaping understandings. As we have seen, maps reflect priorities and biases, whether examined or implicit. Consider something as basic as the choice of map projection (the method adopted for representing Earth's curved surface on a flat map). It is impossible to display the features of a round planet on a flat surface without distortion. As such, the choice of map projection inevitably carries with it a desire to render certain features more accurately than others. For centuries, the dominant projection used in world maps made in North America and Europe was the Mercator Projection, and the Atlantic Ocean was positioned in the center of the map. The projection was useful for navigation purposes, but it introduces huge inaccuracies in the depictions of the comparative size of land areas (those near the poles are greatly enlarged and those near the Equator are shrunk). The result is a map that makes Greenland look larger than Africa, and that, by putting the Atlantic Ocean at the center, pushes East Asia to the periphery.

There is no way to calculate the precise impacts of the widespread use of Atlantic-centered Mercator maps, but they almost certainly helped to foster a North American and European view that marginalized Africa, and to a lesser extent East Asia. In many cases maps are developed and disseminated with the explicit purpose of advancing a political agenda (e.g., maps showing one or another country's territorial claims in a disputed area, such as different names for islands in the sea separating Japan and Korea; maps drawing attention to particular social issues; or maps highlighting the environmental consequences of particular activities). One of the best examples of the link between mapping and politics comes from the Cold War era, when polar projections such as the one shown in plate 5 were created and widely distributed in the United States in an effort to draw attention to the proximity of the Soviet Union and, by extension, to underscore the potential threat it posed to the US.

Technological advances have made it much easier and cheaper to produce maps today than was the case just a few decades ago, and that development alone underscores the importance of nurturing the capacity to think carefully and critically about maps (see chapter 5 for more on this point). Moreover, the proliferation of maps opens up new and important domains for geographical study. How are maps and navigation tools helping or impeding people's efforts to move around and comprehend the surrounding world? What design modifications might make maps more useful for diverse groups of people, including those with physical disabilities? Is the increasingly widespread use of GPS changing the way people think about places and move around in space? What unites all of these new areas of inquiry is the geographical concern with questioning the character, usefulness, and impacts of spatial representations.

Conclusion

More than 200 years ago, the invention of rail transportation paved the way for a massive geographical transformation affecting where people lived, where and how products were produced, how governments controlled territory, and even how individuals thought about distance and what places might be accessible to them. People could move farther from home and stay in contact with their extended families; they were drawn to think about how their lives were intertwined with those living farther away than in the past. Similar dramatic transformations occurred in the wake of the development of the automobile and the airplane. As the twenty-first century unfolds, another revolution in mobility and connectivity is looming – driven not by a single transformative invention but by a suite of technological innovations and social-environmental concerns that are likely to have profound consequences in the years ahead: driverless cars, electric vehicles of various sorts, ride-sharing, super-high-speed trains, and an increasingly pervasive internet. Collectively, these will influence how billions of people experience and comprehend the world around them.

Understanding the implications of these changes and harnessing them for humanity's benefit will not be possible without sustained, thoughtful geographical analysis of evolving spatial arrangements, relationships, and constructs. We need to understand how and why patterns of mobility and connectivity are changing and the variable impacts of those changes on different places and communities. We need to pay attention to how technological innovations are remaking the relationships among places and people – how they are, for example, driving changes in land-use practices in and around cities, differentially affecting segments of the population, altering the connections between rural and urban areas, changing the physical environment, and shifting people's spatial sensibilities. Questions of this sort – geographical questions – are of vital importance in the face of transformations that will likely significantly reshape the organization and experience of life on Earth during the twenty-first century.

3

Places

Geography is often best understood as a discipline defined more by its approach to analysis and explanation than by the objects it studies, but its general concern with the nature of places represents a partial exception. Interest in places is as old as geography itself. In the Western tradition, early Greek geographers were known for their rich accounts of the different places they encountered during their travels, as were their later Arab, Persian, and European counterparts. Moving down to the present, studies of how places are organized and what they look like offer valuable insights into the physical processes and human influences that shaped their development.

George, South Africa, is a modest-sized city lying atop a plateau in the Western Cape region, close to the Indian Ocean. The beauty of its setting and the proximity of forest resources and arable land have drawn people to the area for millennia. George was founded in the early nineteenth century by British Cape Colony leaders. The business district was in the area labeled "City Center" in figure 2.[1] Growth was gradual during the first few decades after the town's founding, but it picked up as transportation connections improved and the timber industry expanded. In the second half of the twentieth century the town became something of a commercial and service hub. Today it is a popular tourist destination and conference center with a population of around 150,000.

Outsiders often think of George as a lovely city with good meeting facilities, nice golf courses, and interesting sites of historic interest. But it is also a place with deep racial divisions. That fact is well known to anyone who spends time there, but looking at the city through a geographical lens reveals much more – including how racial divisions influenced its structure and organization over time, and continue to do so today. In an insightful *Urban Geography* article, Kimberly and David Lanegran point out that the greater George area has been racially segregated since the nineteenth-century founding of the city.[2] A sizeable Colored community (mixed-race, from multiple origins) sprang up a

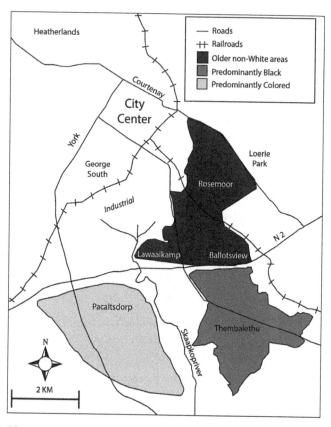

Figure 2 Basic geographical features of the George, South Africa, metropolitan area.

few miles to the south in a place called Pacaltsdorp (fig. 2). Many of Pacaltsdorp's residents worked in George, but Pacaltsdorp had an autonomous political status. As for George itself, the city's sizeable White population was concentrated near the center and an initially small Black population took root in the area labeled Rosemoor on the map. As the Black population grew, it was pushed into Lawaaikamp, and eventually to the south of a major road, the N2, into Thembalethu.

George's geographical character reflects a troubled history of race relations and the impacts of South Africa's apartheid regime between 1948 and 1991. It also served to reinforce segregation in recent years despite the end of apartheid. The George government of the 1990s was anxious to leave the apartheid legacy of the past behind. It was also faced with a housing shortage, and planning for new housing projects was therefore a priority. Land availability and cost factors led most of the new housing projects approved by the government to be constructed on the fringes of mostly Colored Pacaltsdorp and mostly Black Thembalethu. These tended to strengthen, rather than undermine, segregation, however. The presence of the N2 road, which runs to the north of the two settlements, worked against any blurring of the boundaries between these communities and central George. Moreover, even though some of the new housing projects led to Pacaltsdorp and Thembalethu growing physically closer together, that did not foster integration between them because a river separates the two and a bridge linking the two was never built – a product of entrenched ways of thinking that treated the two communities as distinct and fostered urban plans focused on each community's relationship to the city center rather than to each other.

The George example shows that, even in places that experience fairly dramatic changes (in this case the demise of the apartheid regime), what came before matters. If the greater George area had not been so segregated, if the Skaapkop River did not divide Pacaltsdorp and Thembalethu or had been bridged, if the N2 had not obstructed north–south movement, and if the inertia of thinking about Pacaltsdorp and Thembalethu as separate places had not been so great, the post-apartheid story of George would be different. Recognizing the nature and importance of these types of interconnections comes from looking at the geographical organization and character of a place like George holistically. Geographical accounts of places have a role to play in satisfying curiosity, but a focus on place also fosters integrative thinking, raises awareness of the diverse character of the planet, offers insights into questions of identity and belonging, and provides a foundation for critically assessing assumptions about the nature of places and regions. A brief look at each of these shows why geography's concern with the character of places has such an important role to play in the contemporary world.

Place as a Platform for Integrated Thinking

In today's world, taking an integrated approach is championed from the natural sciences through the social sciences and humanities. One important way to approach integrative thinking is to focus on how different factors relate to one another in concrete settings. That is the direction being taken, for example, by researchers interested in the behavior and impacts of fire in mid-latitude forests. No place in those forests has quite the same mix of trees, ferns, lichens, moss, and other types of vegetation. That's because the factors that affect vegetation (e.g., air temperature, rainfall, soil characteristics, exogenous pollution) come together in different ways from place to place. In the process they create distinctive ecological niches, and those niches, in turn, shape fire susceptibility and the ways in which fires or other types of disturbances affect forests. Recognition of the significance of this point has led fire researchers to pay increasing attention to geographical context – a theme that has emerged as a central concern in a recently launched multidisciplinary initiative aimed at "identifying common ground among fire researchers."[3] The underlying idea is that the complex of circumstances found in individual places has a greater impact on fire than had previously been appreciated.

Geographical studies of the impacts of human activities on the evolution of plant communities confirm this general point. University of Kentucky geographer Jonathan Phillips studied how plant communities respond to grazing and fire suppression (the active effort to put out fires in natural areas) in three different settings: central Texas, southwestern Virginia, and eastern North Carolina.[4] His work showed that no single predictive model based on general understandings can capture what has happened. As he put it, "place matters," and the implications of that insight are far reaching. It suggests that instead of assuming we can establish hard-and-fast cause-and-effect relationships and make deterministic predictions, it is more useful to think in terms of probabilistic generalizations and to incorporate the characteristics of different geographical situations into predictive models. To return to the fire example, what this means is that rather than assuming the outbreak of fire in a mid-latitude coniferous forest will have this or that effect, it is more useful to say that if particular sets of conditions are found at certain places within mid-latitude coniferous forests, one or another outcome is more probable if fire strikes.

Place serves as a powerful platform for integrative thinking in the human arena as well. What impact did the expansion of industrial capitalism across Britain in the nineteenth century have on the social relations and activity patterns of women? Bringing a geographical focus on place to bear on that question led Linda McDowell and Doreen Massey to consider the impact of place-based differences between, for example, the Fenlands of East Anglia and the cotton towns of northeast England.[5] They concluded that basic variations in the foundations of these regions' economies, women's role in those economies, the spatial organization of communities, and social norms had fundamentally different consequences for women's roles in more recent times. Just as in the physical geography example, by observing place differences we become aware of the limitations of broad generalizations and recognize the types of contextual factors that shape on-the-ground developments.

The foregoing examples demonstrate the role integrative geographical studies of place can play in deepening understanding of variations across space (a theme also explored in the previous chapter). The lessons that come from such studies are reinforced by growing efforts outside of formal geography circles to consider the impacts of place characteristics on more general issues of concern. Small-scale fisheries are responsible for as much as half of global fish catches, and they play an important role in many regions, especially in less industrialized places. Yet lots of them are threatened by a variety of circumstances, most visibly competition for declining fish stocks because of the harvesting practices of large-scale, commercial fishing operations. Not surprisingly, efforts to promote more sustainable small-scale fisheries have long focused attention on addressing the challenges of overfishing. As important as that issue may be, the sustainability challenges facing small-scale fisheries are more complex. They are tied not just to the number of fish in the ocean but also to ocean health, which is threatened by mounting concentrations of plastics, dying coral reefs, and expanding dead zones (areas where little or no life exists). The vulnerability of fishing communities matters as well – to poverty, disease, hazardous working conditions, and youth unemployment. Recognition of the importance of these latter factors led the Food and Agriculture Organization of the United Nations to promulgate a set of "voluntary guidelines for securing small-scale fisheries" that called for consideration not just of resource management issues, but also of such matters as human rights, respect for cultural diversity, economic sustainability, and gender equality.[6] The overarching message of these guidelines is that holistic, integrated approaches focused on individual places are essential if small-scale fisheries are to remain viable.

A similar kind of message underlies the US National Academies of Sciences, Engineering, and Medicine's recent report *Communities in Action: Pathways to Health Equity.*[7] That report sets forth a conceptual model based on the idea that successful efforts to promote health equity require community-driven approaches that take into consideration the intersecting influences in individual places of the physical environment, income levels, health services, employment, housing, transportation, and education. The Amazon Third Way Initiative provides yet another example. That initiative aims to advance an integrative approach to development that focuses attention on how socio-ecological, technological, and economic characteristics intersect in different places within Amazonia.[8]

As these and countless other examples show, we cannot hope to grasp the complexities of the world around us if we focus solely, or even largely, on discrete topics or objects. Integrative thinking is essential, and places serve as valuable incubators for that kind of thinking. To ask questions about the nature of a place is necessarily to draw attention to the mix of forces – physical, social, economic, political, and cultural – that have shaped it over time. The broadening of horizons accompanying that fundamentally geographical exercise serves as a vital counterbalance to the trend toward hyper-specialization in the modern era.

Encouraging Curiosity About, and Insight into, Earth's Diversity

What led early humans to seek out places beyond their original African homeland and gradually colonize most of the planet? Economic imperatives were a driver in some instances, though in many cases stressed peoples lacked the resources and flexibility to move. Efforts to escape conflict and repression also served as catalysts to movement, but only in limited circumstances. A fuller answer to the question cannot overlook the role of human curiosity, which has always led some people to want to explore what lies beyond the next mountain, the next body of water, the next forest – and sometimes to move there.

That same curiosity continues to drive people to want to learn more about the customs, modes of livelihood, and landscapes found elsewhere. It helps explain why print and electronic media portrayals featuring far-away places attract interest, and why the travel industry has boomed in recent decades. Yet geographical curiosity – and the insights that come with it – should not be taken for granted. Surveys of geographical understanding show great depths of misunderstanding of the wider world. Relatively few people actively seek out information about other places. Most tourists do not venture beyond hotels, resorts, restaurants, and major sites highlighted in guidebooks – often prioritizing venues that resemble the places they came from. For all their horizon-broadening potential, recent technological developments have made many people more sedentary, less likely to explore, and less aware of their surroundings (a subject to which we will return in chapter 5). Globalization has fostered greater interaction across the planet, but that has also fueled resentments and served to deepen the hold of negative stereotypes on people's minds.

Against this backdrop, the value of nurturing geographical curiosity becomes clear. Rich, evocative accounts of different places have an important role to play in this regard. Fortunately, there are a range of people with a geographical bent who are producing such accounts – not just professionally trained geographers but journalists, novelists, travel writers, and film producers. We have a better grasp of the interconnections that make places what they are by reading the work of novelists such as Barbara Kingsolver, James Michener, and George Orwell. Nonfiction writers such as Barry Lopez, Charles Mann, John McPhee, and Andrea Wulf help us understand the past and the present by looking at how human–environment dynamics unfold in geographically sensitive ways. The contributions of these commentators stand in sharp contrast to the many superficial accounts of distant places that play into misunderstandings and prejudices (the prevalence of such narratives points to another reason why we need well-informed geographical commentaries on places).

The last point speaks to a particularly important justification for cultivating geographical curiosity about places: its potential to broaden people's horizons and promote understanding. Many outsiders see Sub-Saharan Africa as a modest-sized, somewhat undifferentiated space dominated by tropical rainforests, poverty, disease, and tribalism (i.e., the usual stereotypes). Viewing the region through a geographical lens, however, reveals an enormous realm of great contrasts – from the tropical rainforests of the Congo Basin to the deserts of Namibia, the East African savanna, and the Mediterranean climate of the Cape; from the remote villages of the West African interior to the vibrant metropolises of Dar es Salaam in Tanzania, Accra in Ghana, and Johannesburg in South Africa; and from the Islamic Afro-Asiatic language speakers of Somalia to the Christian Bantu speakers of western Angola, and the Nilo-Saharan-speaking followers of traditional indigenous religions in South Sudan. Such a view draws attention to the continent's diversity while promoting understanding of Africa's sheer size. Plate 6 offers an important corrective to the diminishment of Africa in the Mercator projection; it shows that the land area of Africa is greater than that of China, India, the United States, and much of Western Europe combined. That this map so often elicits surprise on the part of those who see it for the first time testifies to the importance of focusing attention on the geographical characteristics of places and regions.

Drilling down in scale, geographical portrayals of mountains, valleys, islands, cities, villages, and neighborhoods have great potential to fuel curiosity about the planet and appreciation of its diversity. That curiosity can not only broaden horizons; it can also help to push back the frontiers of understanding by encouraging consideration of the similarities and differences among places. Why have certain less-developed countries such as Nepal and Rwanda been able to achieve better human health outcomes (lower infant mortality rates, longer life expectancies) than other countries that are no higher up on the development spectrum? Why is urban expansion associated with precipitation decline in some cities but not in others? Why have decisions to expand the road network led to significant farmland loss in the United States' Silicon Valley, but not around Bangalore in India?[9]

There are no easy answers to such questions; researchers employing different messages or coming from different theoretical perspectives may well reach different conclusions. But even raising questions of this sort can be insightful – pointing, for example, to the need to consider whether institutional factors have strengthened health care in

Nepal and Rwanda, challenging the assumption that there is a straightforward causal connection between urban sprawl and rainfall, and drawing attention to differences in the relationship between cities and their surroundings in central California and southern India. Curiosity about similarities and differences between places, in short, has the potential to enrich a broad range of inquiries; geography's role in encouraging that habit of mind through its focus on places serves as another reminder of the discipline's significance.

Human Attachments to Places and Regions

People do not simply occupy or visit places; they develop attachments to them that influence what they do, how they think about the world, and even how they construct their own identities. Many, perhaps most, people define themselves at least in part in geographical terms: I'm British, an English person, a Londoner. Moreover, most people do not think about the places where they live, work, and visit in a narrow, mechanistic way; instead they develop a "sense of place" that is as much emotional and intuitive as it is grounded in concrete circumstances. A particular sense of Paris, France, as a place led to land-use regulations in the city proper restricting the height of buildings and promoting conformity with nineteenth-century building styles. In other places housing developments are planned with a view toward creating a space that resonates with people's (often unexamined) geographical sensibilities. Strongly felt emotions about the character of places have stimulated works of literature, music, film, and art, even as they have influenced decisions about where (or whether) to move, plans about where to travel, patterns of support or opposition to development initiatives, and personal decisions affecting the character of the spaces over which individuals or institutions have direct control (gardens, parks, buildings, etc.).

It follows that grasping the nature of places requires consideration not just of their overt characteristics, but also of the way people think about and experience them. In an influential 1976 study entitled *Place and Placelessness*,[10] University of Toronto geographer Edward Relph looked at the proliferation of look-alike commercial strips in and around North American cities. Terming these "placeless landscapes" (because they are ubiquitous and ignore the individual characteristics of the places where they sprang up), Relph invited us to think about what happens when landscapes reflecting people's geographical and historical sensibilities are replaced with cookie-cutter urban developments that can be found anywhere. For Relph, encroaching placelessness undermines the rich diversity of the planet and undermines people's attachment and commitments to the places where they live. Why care much about what happens to a neighborhood if there is nothing distinctive or special about it?

Relph's focus on the emotive and psychological dimensions of place has clear significance for the contemporary world. What leads developers to build vast suburbs with large, stand-alone houses surrounded by water-sucking lawns around desert cities facing water-supply challenges? How does NIMBYism (a "not in my back yard" attitude) figure into the dearth of low-cost housing in some cities or the location of waste dumps in environmentally fragile zones? Taking people's sense of place seriously is a prerequisite to addressing these types of questions.

Moving up in scale, the power of states to shape identity is an issue of fundamental importance in the contemporary world. The British partition of India in 1947 into a Hindu and a Muslim state (India and Pakistan, respectively) divided communities and stoked forms of nationalism that are of great contemporary geopolitical, economic, and social significance. Many of the most intense conflicts of recent decades are rooted in overlapping or discordant geographically grounded identities: Israel–Palestine, eastern Ukraine, Chechnya, Sri Lanka, Sudan, Iran–Iraq, and countless others.

The importance of probing the identity dimensions of places becomes clear when one stops to consider something as basic as the oft-repeated claim that we live in a world of "nation-states" (i.e., states made of a single nation). The word "nation" is one of the more confusing terms in the English language because it has multiple, irreconcilable meanings. Sometimes it is used as a synonym for an independent state (the nation of Indonesia, the United Nations), sometimes it describes a collection of indigenous communities (First Nations), and sometimes it is used to describe substantial ethno-cultural communities seeking states of their own (the Kurdish or the Palestinian nation).

The nation-state idea is rooted in the original meaning of the word: people sharing a common sense of history, culture, and identity who want control over their own affairs in a discrete territory. The French Revolution of the late eighteenth century gave impetus to the idea that the cultural-historical peoples of the world (i.e., the world's nations, in the original sense of the term) should have states of their own. France was formed in the name of the French people, and in the nineteenth and early twentieth centuries other European states emerged out of nationalist movements: a Germany for the Germans, an Italy for the Italians, a Romania for the Romanians, and so forth. But these were far from nation-states in any geographically precise sense. The territories that emerged were not uniformly occupied by people who thought of themselves as French, Germans, Italians, and Romanians because many other peoples lived within these states as well.

It is beyond the scope of this short book to explore the complexities of the relationship between nation and state in detail. What is notable for present purposes is that the nation-state concept was a fiction of sorts from the beginning, and it became even more so as the modern global political map emerged in the wake of the dissolution of Europe's colonial empires. It may be common practice to call the units shown on the world political map nation-states, but the typical state today encompasses multiple cultural-historical communities.

Understanding the nature and significance of the gap between the nation-state idea and on-the-ground circumstances requires critical geographical thinking about the relationship between identity and politically organized territories (formal regions).

That way of thinking reveals the fundamentally aspirational character of the nation-state concept – the hope that the citizens of a given state will think of themselves as a community defined by and committed to a state. This surely is what a Nigerian leader means when purporting to speak on behalf of the Nigerian "nation-state" (given the country's 300-odd ethnic groups [the Hausa, Yoruba, and Igbo are only the largest], extraordinary linguistic diversity [more than 500 recognized languages], and deep divisions between the Islamicized north and the Christianized south).

In reality we live in a world of multi-national states, not nation-states – a seemingly simple, but actually quite profound point that can easily be underappreciated in the absence of critical thinking about the territorial foundations of identity. Practically every day, news from different parts of the world serves as a reminder that the disconnect between the political organization of space and patterns of identity represents one of the profound challenges of the contemporary world. We overlook its nature and significance at our peril.

Questioning Stereotypes About Places and Regions

It is difficult to talk or write about any topic without invoking some kind of spatial construct (a country, a city, an intersection, a biophysical region, etc.). Yet as discussed in chapter 2, the spatial compartments commentators use to frame issues and problems affect how those matters are understood; they hide some things just as they reveal others. That is why it is so important to think critically about the way places are described. When a politician rails against migrants from the Middle East or a journalist makes reference to the crime rate in London's East End or Chicago's South Side, they are encouraging or reinforcing particular ways of thinking about places and regions that can reinforce racial and ethnic stereotypes. As such, it is important to think critically about the appropriateness and usefulness of such framings – a by-product of the geographer's concern with the analysis of places and regions.

Critical thinking about representations of places can also challenge commonly held assumptions about their nature – and the ways they came to be viewed as meaningful spaces. Western Sydney University geographer Kay Anderson made the point powerfully in a classic study of the "Chinatown" district located near the heart of Vancouver, Canada.[11] The city's Chinatown is the area where Chinese immigrants concentrated in the late nineteenth and early twentieth centuries. Interrogating the nature and meaning of Chinatown as a place led Anderson to probe beyond its demographic and architectural characteristics. She showed that the very act of naming the place and treating it as a space apart from its surroundings was a product of the racial ideas and prejudices of Vancouver's dominant White population of European ancestry. Those ideas and prejudices led to government policies and social practices that pushed Chinese people and businesses into Chinatown, limited the opportunities available to the area's residents, and reinforced both a view of the district as an unsanitary, immoral place and on-the-ground circumstances that reflected that view. Thinking comprehensively about the nature of the place prompted Anderson to challenge the prevailing idea that Vancouver's Chinatown was a Chinese invention that stemmed simply from the desire of Chinese migrants to live together in a foreign land.

Turning to a larger-scale example, consider the all-too-common tendency to treat the "Islamic World" as if it were a meaningful geopolitical entity. As popularly used, the "Islamic World" label does not simply reference a part of the world where Islam is the dominant religion; it signifies what is presumed to be an existing, or at least emergent, geopolitical node. Harvard political scientist Samuel Huntington's influential publications on the "clash of civilizations" in the 1990s did much to promote this way of thinking.[12] Huntington argued that the major geopolitical fault-lines of the twentieth century, which he viewed as being political-ideological in nature (communist/authoritarian versus democratic/capitalist), were in the process of being replaced by divisions along cultural-religious lines pitting, for example, the Islamic World against the Judeo-Christian West.

The conflicts that have developed among different peoples living in Southwest Asia and North Africa in recent years have exposed the problem of viewing the Islamic World as even a quasi-unified geopolitical actor. Nonetheless, the Huntington idea still has wide resonance, and its power is easy to see. That power was certainly evident in the wake of the al-Qaeda-inspired attack on the United States on September 11, 2001. Iraq became a central focus of retaliation even though Saddam Hussein and al-Qaeda were fundamentally at odds. Iran and Iraq were preposterously lumped together into an "Axis of Evil" even though the two countries had fought an eight-year-long bloody war with one another just over a decade earlier. One of the justifications advanced for intervention in Iraq was to head off the establishment of "a radical Islamic empire that spans from Spain to Indonesia"[13] – an argument that assumes the region has enough in common socially and culturally for the emergence of a unified empire to be a conceivable reality.

These ways of thinking have not disappeared. Retired US Brigadier General and periodic Fox News commentator Nick Halley (the author of the provocatively entitled book *Terrorism: The Target is You! The War Against Radical Islam*[14]) is on the lecture circuit giving talks arguing that the Islamic World represents an existential threat to the rest of the planet. It would be naïve not to recognize that there are certain extremists who, in the name of Islam, have expansionist goals and view violence as a legitimate means of pursuing them (though almost no one would give credence to Halley's claim that their numbers exceed 100 million). But how can one consider the prospects of these extremists achieving such goals without some appreciation of the obstacles that exist to creating a united front in the part of the world where Islam is the dominant religion? How can one possibly begin to understand the "Islamic World" if representations of its status as

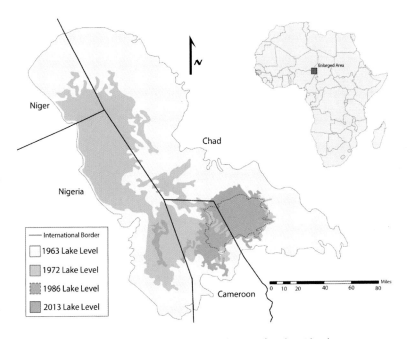

Plate 1: The dramatic shrinkage of Lake Chad.

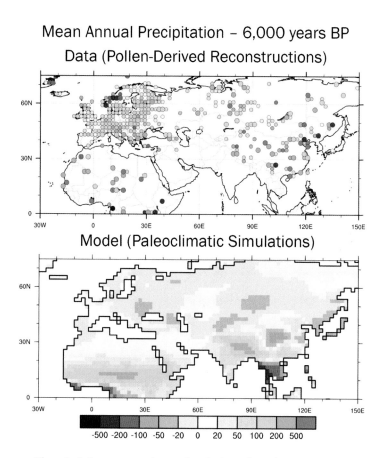

Plate 2: Map comparisons that help refine climate models.

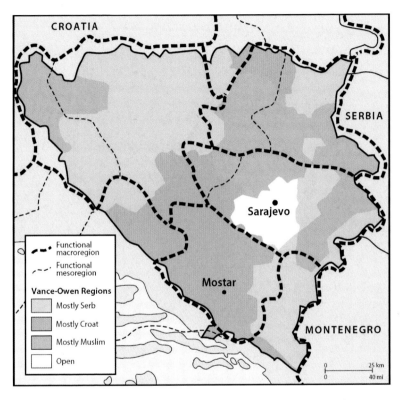

Plate 3: Geographic analysis of a proposed partition plan for Bosnia-Herzegovina in 1993.

Plate 4: Global map of travel time to cities in 2015.

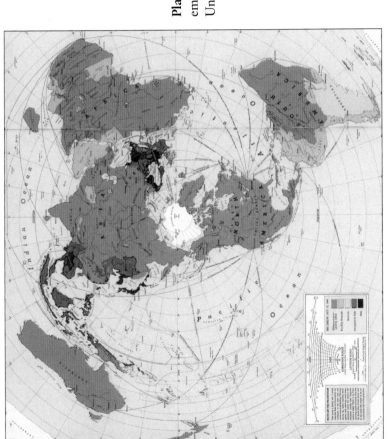

Plate 5: 1944 polar projection emphasizing the proximity of the United States and the Soviet Union.

Plate 6: Visualization showing Africa's true size.

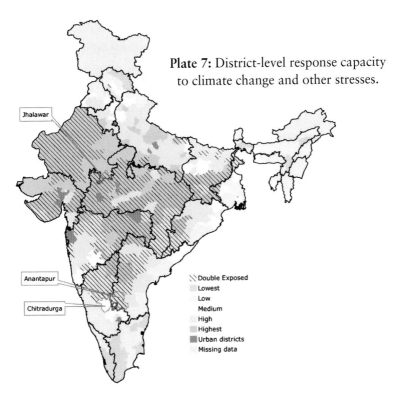

Plate 7: District-level response capacity to climate change and other stresses.

Jhalawar

Anantapur

Chitradurga

Double Exposed
Lowest
Low
Medium
High
Highest
Urban districts
Missing data

A Divide In Culture And Politics

BELARUS

POLAND

Kiev

Lviv

Striped areas were won by the opposition in the 2010 presidential election

SLOVAKIA

UKRAINE

Kharkiv

RUSSIA

Luhansk

Won by Viktor F. Yanukovych, now Ukraine's president

Percentage of population that natively speaks **Russian**:
100% 80 60 40 20 0

Ukraine's political split reflects a deeper cultural divide in the country. In the 2010 presidential election, the opposition won in all of Ukraine's western provinces, where most people speak Ukrainian rather than Russian and many call for deeper economic and political ties with Europe.

MOLDOVA

Odessa

ROMANIA

BLACK SEA

RUSSIA

Bucharest

Sevastopol

150 MILES

Sources: State Statistics Service of Ukraine, Ukraine's Central Election Commission

THE NEW YORK TIMES

Plate 8: The political and ethnolinguistic situation in Ukraine in early 2014.

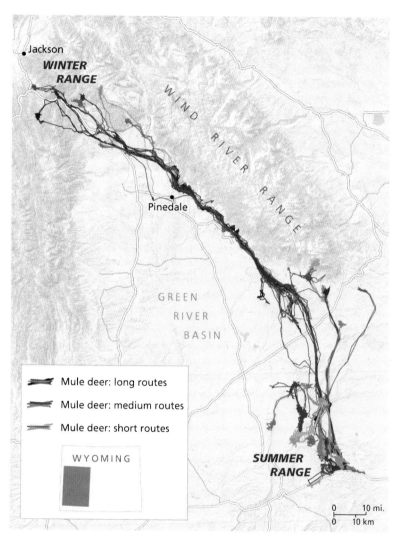

Plate 9: Mule deer migration corridors in western Wyoming.

a singular geopolitical node are left unchallenged? What is termed the Islamic World is in fact riven by deep-seated differences, not just over doctrinal matters related to succession to the caliphate (the original source of the Shiite–Sunni split), but also over cultural practices, modes of livelihood, political ideologies, and nationalist affiliations. Recognition of such geographical basics is a precondition for any kind of thoughtful consideration of developments in Southwest Asia and North Africa, irrespective of the political views one brings to the table.

Few examples are as stark as the one outlined above, but critical thinking about regional representations is the only way to expose problematic assumptions and challenge troubling stereotypes about continuities across space – overarching Western stereotypes that cast all of Mexico as dangerous, Sub-Saharan Africa as disease-ridden, northern Canada as pristine, and inner-city Detroit as violent. Critical geographical thinking about places, in short, represents the best line of defense against unconsidered or deliberately manipulative representations of different parts of Earth's surface.

Conclusion

In 2007, Linda Lobao and two of her colleagues argued that sociological research on inequality was overly preoccupied with the national scale and gave too little attention to contextual circumstances.[15] Their concern with understanding place-based influences – one of geography's core concerns – applies in many other arenas as well. Where something happens usually affects what happens. An effort to promote economic development by offering tax breaks to lure business investment may have positive socio-economic consequences in one place, but negative ones in another depending on the mix of existing businesses in nearby communities, local employment opportunities and labor conditions, and the attitudes of the local community. Cattle grazing in stream banks will have different impacts depending on the vegetation, soils, topography, and stream flow at the place where the grazing is taking place. The point is that geographical context matters. The more we pay attention to place-based influences on physical and human processes, the better we will be able to appreciate the nature and significance of the planet's underlying diversity.

4

Nature and Society

Some years ago I participated in a conversation with a group of colleagues at my university about the formation of a new interdisciplinary program focused on the German-speaking part of Europe. A number of excellent suggestions were made about the curricular components for such a program: courses on German language and literature, history, economics, politics, philosophy, and music. Until I spoke up, however, not a single person made any reference to the natural environment, land-use, or ecological challenges. Although my suggestion to consider such matters was immediately embraced, it was fascinating to note that, before my intervention, a group of highly intelligent, educated people were able to think about what it takes to understand a part of the planet without considering the physical environment or the relationship of humans to it.

In today's world, human-driven environmental change is receiving unprecedented attention, yet the tendency to treat humanity and the natural world as discrete realms remains strong. Modern infrastructure allows most urban dwellers to separate themselves from the environment in their day-to-day lives (except in extreme cases, a major storm is little more than a minor inconvenience). Nature is treated as something one goes out into and experiences on the occasional foray out of town rather than as something integral to existence. Most universities have separate administrative divisions for the sciences, social sciences, and humanities – reinforced by discrete deans and protocols for teaching and research. In primary and secondary schools, taking students on field trips is far less common than it once was, and the classroom is increasingly insulated from the outside world. As BBC radio producer and author Tim Dee lamented on the pages of the *New York Times* a few years ago, "Nowadays nature tables [tables where pupils bring in natural items they've discovered outside for display and discussion] are not welcome in schools. They are deemed dirty, perhaps dangerous, and potentially illegal."[1]

Against this backdrop, it is not surprising that, outside of discussions of pressing environmental problems, relatively little attention is given to the interrelationship between society and nature. By thinking geographically, we can challenge this tendency. Trying to understand the variable character of Earth's surface offers insight into both physical and human influences. Focusing attention on the character of a place such as Venice, Italy, invites consideration of why people came to occupy the city's improbable site in the middle of a lagoon; how the hydrologic, geomorphic, cultural, and socio-economic features of the place shaped its development; and what physical and human challenges lie ahead. And the city's landscape – the character and organization of its building, streets, canals, squares, and walkways – can be read like a book, telling us stories about the conjunction of forces that produced historical and contemporary Venice.

It is challenging to grapple with the nature–society dynamic because each side of the equation is complex, and there are great differences in the theories and methods appropriate to understanding the physical and human world. A failure to embrace these complexities prompted some early twentieth-century scholars, including professional geographers, to embrace the idea that environmental context determines cultural and social outcomes (an approach labeled environmental determinism). That way of thinking led to some alarmingly simplistic, ahistorical, often racist views of the world – assertions, for example, that the people of the tropics are indolent and incapable of great achievements because the environment made them so. As geographers and others came to think more deeply about nature–society linkages, most of them abandoned that way of thinking, though it continues to crop up with some regularity among those who have not been exposed to geographical scholarship over the past seventy-five years (an argument in and of itself for the importance of studying geography).[2]

In the wake of the rejection of environmental determinism, many geographers began focusing solely on either the physical or the human world. Nonetheless, a realm of inquiry fundamentally concerned with the nature of places and with the character and consequences of different geographical arrangements cannot ignore nature–society relationships for long, and geographical work on the subject has blossomed in recent decades. Indeed, of the traditional disciplines, geography today is the one that arguably is most centrally concerned with looking at the interrelations among and between natural and human processes on Earth's surface.

To be sure, many scientists and social scientists

grapple with questions that cross the human–physical divide. To cite but a few examples, environmental chemists look at the impacts of human-caused pollution on water quality, forest ecologists seek to understand how human actions affect the diversity of flora and fauna in forests, and environmentally oriented legal scholars look for ways to design rules and regulations that can curtail emissions from the burning of fossil fuels. These and other efforts to bridge the human–environment divide do not make geographical inquiry irrelevant, however, because geography offers perspectives and techniques that provide insight into the spatial and material character of nature–society interactions as they play out on Earth's surface, even as it encourages integrative thinking about places and ecosystems.

Three distinct, yet overlapping, concerns are at the heart of geography's approach to the human–environment dynamic. One strand focuses on the insights that come from the study of distributions and patterns of relevance to nature–society relations – a strand rooted in geography's spatial tradition, discussed in chapter 2. A second derives from an interest in how the characteristics of particular settings or places influence nature–society interactions – an outgrowth of the geographical concern with place, discussed in chapter 3. An interest in how human–environment transformations in one place influence, and are affected by, developments in other places represents a third characteristic of geographical work at the nature–society interface. That concern also draws attention to how the space or the scale that is used to frame issues influences understandings.

Studying Distributions and Patterns

In the face of extraordinary droughts in South Africa and Mongolia, swamped shorelines in the Maldives, record-breaking temperatures throughout much of the world, and rapidly retreating ice sheets near the poles, one of the signal challenges of our time is to figure out how climate change will affect different parts of Earth's surface. Maps showing which places are more or less vulnerable to climate change based on understandings of the climate system are useful, as they provide insight into areas where flooding, drought, and temperature extremes present greater threats. These maps only scratch the surface of what a geographical concern with distributions and patterns can tell us, however, because climate-change vulnerability is not simply a product of physical-environment change; the character of human societies and associated institutions matters as well.

When viewed against this backdrop, the importance of geography's deep engagement with spatial variability becomes clear. Some places are in a much better position to cope with the consequences of climate change than others. They have a greater capacity for resilience because they are wealthier; they have flexible, diversified economies that can cope well with stress, their governmental institutions function efficiently and are respected, and existing infrastructure gives them a head start. Taking these types of factors into consideration is critical if we are to understand the vulnerability of different places to climate change. That is what geography professors Karen O'Brien at the University of Oslo and Robin Leichenko at Rutgers University did (in collaboration with others) in their study of the vulnerability of different agricultural communities to climate change in India.[3] They examined the distribution of biophysical, economic, social, and technological factors, and then constructed a composite map (plate 7) that shows how the intersection of these factors will likely affect patterns of vulnerability there. The map takes into consideration, for example, the greater coping capacity of areas where many different crops are grown under diverse conditions as opposed to those that, under the pressure of neo-liberal reforms, have come to rely on the large-scale, export-oriented farming of a single crop where a change affecting the dominant crop could have devastating impacts. It also shows areas that face a "double exposure" problem because of the additional vulnerability of their agricultural economies to import competition.

The kind of analysis undertaken in this study underscores the value of geography's deep concern with distributions and pattern. Virtually all threats to biodiversity and ecosystem loss exhibit distinctive, often revealing, spatial patterns. Analyzing the diffusion of pollutants across geographical space, for example, is essential to understanding their impacts on natural systems. The analysis of mapped data is foundational to most efforts to predict areas of vulnerability to drought and hunger, understand the causes of flooding, and assess the impacts of forest management practices. Studying changes in the distribution of selected flora across space yields insights into the impacts of climate change, forestry practices, the use of herbicides, and much more.

Mapping, however, is not simply an end product. What makes the study of distributions and patterns really useful is careful, creative thinking about what should be mapped, how phenomena might be mapped, and what to make of the relationships revealed by such efforts. That is where the importance of studying geography becomes clear, for geographical training is not just about becoming familiar with mapping software; it is about learning to ask good questions about patterns and distributions. Asking such questions not only offers insights into applied challenges (e.g., where and how to construct roads to minimize impacts on wildlife migrations); it can also promote consideration of neglected but important human–environment interconnections. That type of inquiry is what led scholars and activists to explore the heightened challenges many minority communities face as a result of policies and activities that expose them to elevated environmental threats. Concerns over such matters have made commentators with geographical instincts important contributors to the "environmental justice" movement – a movement concerned with addressing the unequal distribution of environmental benefits and burdens across different communities as a consequence of racist or classist attitudes and practices.[4] Geographically informed assessments of the environmental challenges facing marginalized communities have helped draw attention to the environmental dimensions of discrimination, and the study of environmental justice is now part of the geography curriculum in many colleges and universities.

The study of geographical patterns and distributions has become such an important area in the study of human–environment relations that it has given rise to major interdisciplinary research initiatives, such as the Land Change Science program in the United States.[5] As described on the website of the US Geological Survey:

> The surface of the earth is a patchwork mosaic of natural and cultural landscapes. Each of these patches is part of a very diverse and interconnected spectrum of landscapes ranging from relatively pristine natural ecosystems to completely human-dominated urban and industrial areas. The mosaic is not static, but regularly shifts due to changes resulting from natural phenomena and human activities. In an effort to better understand these changes and their associated impacts a new field of study has emerged called Land Change Science.[6]

Geographers have played a leading role in the Land Change Science initiative; it serves as a telling example of the contributions the geographical analysis of patterns and distributions can make to the effort to understand nature–society relations.

How Places Influence Human – Environment Interactions

Many of the scientific and technological break-throughs of the last century have come from people immersing themselves in particular problems – how to store energy in a battery, how to sequence the human genome, and the like. The trend toward specialization can have a downside, however; it can work against the kind of more comprehensive thinking that is needed to address complex human–environment challenges. Focusing attention on the nature and importance of geographical context – of places – has an important role to play in promoting that type of thinking.

Gilbert White's pioneering work on human settlement in and around floodplains in the United States is a case in point.[7] During the early decades of the twentieth century, the management of flood-plains was seen largely as an engineering problem; as a result, a growing number of dams and levees were built so that more people could settle in floodplains. White's geographical bent led him to question the wisdom of an approach that treated floodplain management merely as a technological challenge. For him, floodplains were complex systems that required a management approach focused on accommodation and adaptation. He thus championed policies that discouraged settlement in highly vulnerable locations, and instead focused on a balance between human and natural factors. In the process, he helped open up a new field of study, hazards geography, which today is a thriving geographical subdiscipline. Sadly, it seems to take the experiences of places such as New Orleans in the wake of Hurricane Katrina to demonstrate the wisdom of White's approach to floodplain management – and even then, its lessons are often quickly forgotten.

The importance of the kind of holistic geographical thinking championed by White extends into many other facets of the nature–society dynamic. Wrestling with the issue of soil erosion in developing countries in the early 1980s, University of East Anglia geographer Piers Blaikie began questioning the dominant wisdom that soil loss was attributable solely to mismanagement, overpopulation, or shifting ecological circumstances. Looking at a Nepalese case study through a geographical lens, Blaikie came to realize that other factors were at play – most obviously economic and political pressures that were pushing poor farmers onto steeper slopes that could not be cultivated without exposing soils to severe erosion.[8] Blaikie's work served as a catalyst for the further development of the emerging field of political ecology and encouraged those with interests in the subject to consider how the local and regional expressions of political-economic arrangements shape ecological outcomes.

Over the past three decades, studies of a variety of human – environment issues have demonstrated the importance of taking geographical context seriously. University of Arizona geographer Diana Liverman, for example, set out to see if local conditions played a significant role in farmers' struggles to cope with drought in the Mexican states of Sonora and Puebla.[9] She found that variations in rainfall deficits could not adequately explain why agricultural losses were greater in some areas than others; instead she concluded that losses were more often a consequence of lack of access to irrigation technology and of land-tenure arrangements that gave local farmers little control over how land was used or how land-use decisions were made. More recent work in other places has confirmed the broader applicability of this insight.

Studies along these lines consistently show that the vulnerability of farm populations and food production systems to climate change, economic shocks, and blights cannot be adequately explained without reference to the mix of institutional and economic arrangements, social circumstances, and cultural norms found in different places.[10] They thus call into question overly simplistic generalizations, such as the assumption that climate-change stress is directly responsible for conflict in less industrialized parts of the world. A multi-year study of armed struggles in Sub-Saharan Africa by University of Colorado geographer John O'Loughlin and colleagues exposed the fallacy of that assumption; they showed that conflicts in the region were more strongly associated with local economic and political factors than with climate-related stresses.[11]

Taking geographical context seriously can also draw attention to human practices that work against environmental sustainability. Throughout much of their histories, cities such as Phoenix, Arizona, and Doha, Qatar, developed with minimal reference to the surrounding physical environment. Greater contemporary sensitivity to environmental matters has led to some efforts to shift course, but in many other places humanity is only beginning to wrestle seriously with the disjunctions that exist between urban form and bio-physical context. Today the urban landscapes of a plethora of new and rapidly growing cities all over the world reflect similar visions of what a modern world city is supposed to look like and similar ideas about the types of urban developments that symbolize power and importance. The same architects, structures, and landscape designs pop up over and over again, no matter what the local physical-environmental setting might be.

The ideas that have influenced the development of these cities, and their landscape outcomes, are essentially placeless, in the sense that the previously discussed cultural geographer Edward Relph used the term (i.e., they are divorced from local context). The result is urban developments that are poorly attuned to the surrounding environment: consider the large grass lawns around Phoenix, Arizona, the monumental fountains in Astana, Kazakhstan, and the lush islands off the coast of Dubai – all of them at odds with their desert setting. It will take new ways of thinking – ones rooted in geography's deep engagement with place – as well as accompanying shifts in political will, to change course in any meaningful way.

Interconnections across Space and Scale

In the face of concerns about declining crop yields in a drought-stressed part of South Korea a few years ago, the Daewoo Group, one of the nation's major conglomerates, entered into a lease with Madagascar's government to secure shipments of grain that took almost half of that island's arable lands to produce.[12] Uprisings in Madagascar over the terms of the deal proved to be enormously disruptive and likely played a role in the ousting of the country's president. A few years later, growing awareness of the health benefits of walnuts in China led to a surge in demand for the product, which in turn prompted significant land-use changes in the northern part of California's Central Valley as fields devoted to other crops and undeveloped land were converted into walnut orchards – with implications for water use, soil erosion, and labor needs.

These and countless other examples show that understanding human–environment dynamics requires consideration of connections across geographical space. What happens in one place is very often affected by, and affects, what happens elsewhere. The importance of tracing those connections provides another reminder of the value of geographical thinking, given its emphasis on situating individual events and circumstances within a wider web of conditions and occurrences. It seems straightforward to view the surge in greenhouse-gas emissions in China over the past twenty years as the product of a single-minded internal development strategy; it is also important to understand, however, how the outsourcing of production by Western countries to China, together with an accompanying shift in the location of pollution-generating industries, has made it a carbon-emissions hotspot.[13]

Tracing and analyzing connections across space is certainly critical to any serious effort to confront environmental sustainability challenges. Consider the local organic foods movement that has taken root in the United States and Europe over the past couple of decades. The movement was driven by a desire to promote sustainable local farms, undermine the power of corporate agriculture, and reduce the environmental impacts of the long-distance transport of consumables. Despite its laudable objectives, a certain geographical myopia became apparent when followers of the movement began calling on governments to restrict organic produce coming from developing countries. That was the approach taken some years ago by the Soil Association, a UK-based group devoted to championing organic, sustainable farming in the country. An opinion piece in the *San Francisco Chronicle* written by Macalester College geographer William Moseley drew attention to an easily overlooked issue, however.[14] If Europeans and North Americans stop buying organic products coming from Africa,

Mexico, and Central America, organic farmers in those regions will have little choice but to return to producing pesticide- and herbicide-intensive staple export crops, with seriously negative environmental and social consequences. If connections such as these are not even considered, it is impossible to devise policies that result in net-positive benefits for people and the environment.

Geography's concern with human–environment matters involves exploring the networks and flows that connect places together and investigating how production and consumption practices in one place influence human–environment dynamics in others. It also means bringing a critical perspective to bear on the spaces and scales that are used to frame analyses and interpretations of nature–society dynamics. This latter issue is important because unexamined geographical framings of human–environment issues are commonplace. Countless articles and commentaries make reference to China's air pollution problems, population pressures in Sub-Saharan Africa, biodiversity loss in Amazonia, Alaska's shrinking glaciers, and the role of "the Global South" in global climate negotiations. Yet each of these carries with it a particular (often unexamined) spatial delimitation of an issue that can hide as much as it reveals. Air pollution does not stop at China's borders, its causes are not simply Chinese, and the extent of the pollution problem varies greatly within the country. A few glaciers in Alaska are advancing, even as most are retreating, and the former does not refute the case for a warming planet if the focus is physical processes beyond the confines of Alaska. Aside from being a geographically inaccurate term that carries with it a sense that a direction on the compass is associated with a particular set of socio-political and economic characteristics (a framing with environmental determinist connotations), the "Global South" is hugely diverse; it is comprised of countries, regions, and cities with highly variable stakes in climate-change negotiations.[15]

If we pay heed to our geographical assumptions, it becomes easier to see that what makes social and environmental sense in one setting doesn't necessarily make sense in another. Urban agriculture is a common practice in many cities in less industrialized parts of the world, where it has multiple advantages. In the case of Dar es Salaam, Tanzania, for example, large urban farms provide green, cool spaces where people socialize, exchange news, and engage in creative problem-solving; they are sites of political rallies; they are areas otherwise unsuitable for other types of development where healthy, nutritious fruits and vegetables are grown; they are sources of employment and income for farmers and those who set up adjacent businesses catering to farm-product customers (lunch shacks, small retail stands, etc.); they provide routes through the city that are widely thought to be safer from crime; and they foster independence and a sense of pride among those who cultivate them.[16]

Nonetheless, cities such as Dar es Salaam have faced significant efforts to reduce the amount of land they devote to the cultivation of fruits and vegetables. Those efforts reflect the influence of a Euro-American-derived idea that large-scale urban agriculture is a sign of backwardness – an obstacle to the development of a modern, prosperous city. Even Westerners who resist this logic often look askance at the farms in Dar es Salaam because they do not have the ordered, tidy appearance of the lots they are used to seeing. In the absence of any serious consideration of the appropriateness of relying on urbanization norms developed in one part of the world to guide planning decisions in another – the type of thinking the study of geography encourages – it is all too easy to overlook important ways in which the importation of norms from other places works against local interests.

In a related vein, efforts to promote socio-economic development and environmental sustainability in poorer parts of the world often fail because generalized ideas about how to intervene effectively come up against local circumstances that create unintended consequences. To cite a much-publicized example, the Millennium Villages Project championed by Columbia University Professor Jeffrey Sachs led to an influx of cash and infrastructure in Dertu, Kenya (among other villages). After a promising start, however, the project ended up undercutting traditional arrangements and ways of doing things – drawing many new settlers into the town and undermining its role as a nomadic stopover. The results were described in an interview with Nina Munk, the author of *The Idealist: Jeffrey Sachs and the Quest to End Poverty*:

> They were now really living in a kind of squalor that I hadn't seen on my first visit. Their huts were jammed together; they were patched with those horrible polyurethane bags that one sees all over Africa. … There were streams of slop that were going down between these tightly packed huts. And the latrines had overflowed or were clogged. And no one was able to agree on whose job it was to maintain them. And there were ditches piled high with garbage. And it was just – it made my heart just sink.[17]

In our increasingly interconnected world, places are affecting one another in ways that have profound social and environmental impacts. Understanding those impacts requires taking geographical variability seriously, as well as paying careful attention to the insights that come from people living in different places. There is a certain irony to the latter point because geography long marginalized local knowledge. Yet as geographical thinking about place and context has matured, the problems of that stance have become increasingly apparent – giving rise to a modern discipline that, along with anthropology, is among the most attuned to the value and insights that come from local knowledge. No serious encounter with the social and environmental complexities of the planet is possible without it.

Conclusion

The rigid lines that often separate the natural sciences from the social sciences and the humanities – lines that are sustained by separate colleges of natural and social science, training programs that rarely cross the human–physical divide, and the inertia of research cultures within disciplines that see themselves as being on one or the other side of the divide – mean that cross-cutting work is all too rare. The situation is changing in the face of rapidly mounting evidence of the rate and extent to which humans are altering the environment on which we depend for our existence. Given the magnitude of the challenge, though, efforts coming from many different backgrounds and perspectives are needed to address what is happening.

Modern geography is just one of these, but it is an important one. Its concern with spatial arrangements offers an approach to organizing, displaying, and analyzing information that sheds light on what is happening; its place-based concerns promote thinking about important interconnections that cross the human–physical divide; its concern with context offers insights into the ways in which processes are affected by the circumstances present in specific locations; and its insistence on raising critical questions about how, and at what scale, we divide up the world focuses attention on the advantages and limitations of different approaches to nature–society questions. To think geographically is to be open to a range of forces, biophysical and human, that shape, and are shaped by, the places and spaces that make up the planet. That mode of thinking is desperately needed in a world that is more and more threatened by an increasingly dominating and rapacious human presence.

5

Why We All Need Geography

Geography's value extends beyond the substantive and analytical contributions it makes to research, policy-making, and planning. The subject also has a fundamentally important role to play in creating an informed, engaged, enriched populace. Some students who pursue advanced geographical study end up teaching in primary or secondary schools, or in universities. Others draw on the ideas, perspectives, and skills they learn as students of geography to reach broader audiences. Yet others simply find that studying geography makes them more aware of their place in the world, more alive to its complexity, and more curious about other people and places. No matter how it is accomplished, geographical education is vitally important, for exposure to geography and associated ways of thinking can open people's eyes to the richness and diversity of the wider world, contribute to the development of an informed citizenry, and provide insights into present conditions and future possibilities.

Consider what is lost if geography is not part of the educational mix. Students may never be encouraged to develop even a basic understanding of how the world is organized environmentally, politically, and culturally. They may never be encouraged to think of the landscape as a window into human and physical processes. They may never be challenged to think about how the spatial organization and material character of the places where they live are similar to, or different from, other places. They may never gain much insight into the potentials and limitations of increasingly ubiquitous geospatial technologies (GPS, GIS, internet mapping applications, remote sensing). They may never learn to think about how maps can be used to convey, or distort, information. And they may never be asked to think about the deep interconnections between nature and society – or have the opportunity to develop the intellectual tools to understand or challenge political claims or policy proposals related to the environment.

To be sure, appreciating what geography has to offer requires moving beyond a shallow place-name-based conception of the subject that remains distressingly common among the general public – and even in the political arena, where geography's concerns with difference and diversity introduce complexities that are sometimes deliberately pushed to the side. Making a compelling case for the importance of geographical education is not hard, however, given the subject's potential to promote awareness of the wider world, enhance people's lives, strengthen civil society and policy-making, and facilitate understanding and use of increasingly prevalent geospatial technologies.

Promoting Awareness of the Wider World

In our present hyper-connected era, most people know something about other places. Yet how much is known is open to serious question. From a geographical perspective, the most visible edge of the problem can be seen in stories of people assuming that Africa is a country, thinking that Thai people come from Taiwan, or not knowing the Andes are in South America. A well-publicized 2006 National Geographic–Roper survey in the United States revealed that only one in ten 18- to 24-year-olds could place Afghanistan correctly on a world map, and nearly half thought Sudan was located in Asia.[1]

(Sadly, respondents from the United Kingdom and Canada did not do much better, even though geography occupies a more prominent place in the curriculum in those countries.) Even the oft-quoted statement (of unknown origin) that "war is God's way of teaching Americans geography" is not holding up. In the same survey 63 percent of young Americans could not find Iraq on a map despite the fact that the United States had invaded the country three years earlier and Iraq was still in the news on a daily basis.

There is something self-evidently amiss when a significant percentage of the population in countries with the global reach of a United States or a United Kingdom has little grasp of basic locational facts. Indeed, it is impossible even to begin to have a serious discussion about what is happening in the world without some basic place-name knowledge. Yet reducing the geographical ignorance problem down to location facts risks reinforcing the simplistic place-name view of the subject noted earlier. Indeed, knowing location facts may be less important to global geographical awareness than knowing that more Muslims live in Indonesia than in any other country, that Amazonia plays host to the greatest biodiversity on Earth, that the European Union is the United States' largest trade partner, that warmer

113

water can make hurricanes more intense, that melting ice sheets raise sea levels, that when it is summer in South Africa it is winter in Europe, or that there are few ethnically homogeneous countries (to mention just a few examples).

The handful of geographical fundamentals listed above might appear to be little more than trivia, but they are just as important to understanding the human and environmental context in which humanity is embedded as knowing basic historical facts is to making sense of the evolution of the human story. Having no appreciation of Earth's environmental, social, political, and cultural make-up in our interconnected age is like living in a bedroom on the ground floor of a house without any sense of the overall size and shape of the structure, what the rooms upstairs are like, or how they are arranged relative to one another. Some grounding in geographical patterns and processes, in short, is critical to making sense of the planet we occupy and to putting developments and events into context.

Growing geographical awareness can also play an important role in raising expectations of what the public has the right to expect from media reports and policy statements addressing developments around the world. In the immediate aftermath of the conflict that broke out in Ukraine in early 2014, few commentaries offered significant insights into relevant historical or ethno-territorial circumstances. One exception was a *New York Times International* story, which was accompanied by the map shown in plate 8. That map, assembled by a graphics editor with a significant geographical background (Derek Watkins), offered a set of telling insights into the divisions wracking the country: the east–west split between native-Ukrainian-speaking areas and those dominated by native Russian speakers, the relationship between that split and the election that put the pro-Russian leader Victor Yanukovych in power, and Ukraine's situation in relation to other countries. In the absence of a general population that is able to appreciate the kinds of insights that come from such a map, and has the ability to understand and interpret it, the incentive to produce such cartographic products will disappear. A geographically educated populace, then, has a role to play in fighting the slide toward increasingly superficial assessments of complex developments around the planet.[2]

Exposure to geography can also serve to arouse interest in, and curiosity about, other peoples, places, and landscapes. There are countless stories of children's imaginations being stirred after looking at an atlas or reading a geographical portrait of another place. As these children's eyes are opened to geographical differences in environment, culture, society, and economy, many come to develop a growing appreciation for the planet's diversity and an expanding grasp of their place in the larger mix. They want to know more, and they begin to understand how and why the world looks different when viewed from elsewhere.

It is difficult to overstate the value of expanding geographical curiosity. Not only does it encourage individuals to fill out their mental pictures of Earth's physical and human building blocks; it also heightens their appreciation of difference while reducing their tendency to embrace problematic stereotypes about other people and places. It gives them the tools to think about how the world looks when viewed from elsewhere. And it sensitizes them to the resiliency and fragility of communities and environments.

Knowing about something is a prerequisite to caring about it. If the Amazon or Afghanistan means nothing, how can we expect people to be concerned about deforestation in the former or endemic conflict in the latter? Most of the world's peoples are deeply entangled with other parts of the globe in their daily lives – from the food they eat and the clothes they wear to the sites they access on their computers to the very air they breathe. Giving form and substance to peoples and places in other parts of the world makes it harder to view them as if they were video-game depictions – devoid of natural and built wonders or thoughtful, interesting, flesh-and-blood people. Against this backdrop, geographical awareness is more than a luxury; it is essential to living thoughtful, caring, responsible lives.

Enriching People's Lives

Exposure to geography has personal significance as well; it has the potential to connect people more closely with their surroundings and give greater meaning to their lives. Throughout the world, a growing number of people are spending more and more time on their cell phones, working and amusing themselves on their computers, playing video games, and streaming videos. Some of these can promote awareness and understanding of the wider world, but a brilliant Gary Varvel cartoon speaks to the downside: after being taken outdoors, a young boy declares, "Oh, I've seen this level on my video games."[3] The technological revolution of the past few decades has come at the expense of direct experience with the tangible environment and face-to-face contact with other people. We are only beginning to understand the consequences of these developments, but there is already growing evidence of their negative effects: dampened sensitivity to the look and smell of the environment, reduced adventurousness, and greater loneliness and depression.

Where, in today's educational world, does a student find encouragement to climb a tree, look around, and wonder about the planet and its place in the universe? Where is the stimulus to marvel at the mix of plants growing in a nearby meadow, to poke around back streets and alleys, to explore beyond the usual tourist sites during a vacation, to wade into a stream, to look out the window of an airplane instead of immediately closing the shade, or to spend time in a neighborhood or place that challenges ingrained comfort zones? To be sure, some people will always be driven to do these things, but the available evidence suggests that their numbers are declining.[4]

Geographical education alone cannot reverse this trend, but it can have a salutary effect. Many geography courses devote considerable attention to what the landscape can tell us about physical and human processes, and some incorporate field trips aimed at heightening students' observational skills and enhancing their appreciation of the character of the material environment. Introducing students to cartography, GIS, and remote sensing courses can encourage thinking about, and explorations of, physical and human patterns. Studying physical geography can stimulate interest in, and understanding of, everything from why there is wind to the forces that shape the topography of a place. Explorations of human geographical topics can foster curiosity about cultural patterns, the organization of cities, and the landscape impacts of economic and social processes. Coursework emphasizing human–environment relations can promote awareness of matters ranging from how environmental attitudes influence land-use decisions to the causes and potential consequences of building in flood zones. Geographical studies of faraway regions can spark interest in other places and prompt students to see for themselves what life is like elsewhere.

The point is that geography, when taught well, has the capacity to challenge physical and mental bubbles that constrain thinking and experience. It also can enrich people's lives, much in the way the study of philosophy, history, or literature often does. As the ancient Greeks (among others) made clear, the purpose of education is not simply to impart practical skills (though geography offers plenty of these); it is also to enhance people's intellectual, social, and psychological well-being – to enrich their minds by fostering curiosity, awareness, and appreciation for things they might otherwise take for granted. You can walk a street and see it as nothing more than a path; alternatively, you can look around and think about how and why the landscape looks the way it does, contemplate the character of the buildings, and ponder the mix of people who are present. You can simply turn on a tap to get water, or you can reflect on where the water came from and whether current levels of water use are sustainable. You can look down a mountain valley and admire the beautiful scene, or the experience can also prompt you to think about how the valley came to be, why the mix of vegetation is different on one side as opposed to the other, and how the valley's particular location in relation to other physical features and human patterns might have facilitated or impeded the migration of people.

Developing an understanding of and appreciation for geography helps move people to the second alternative in each of the foregoing examples. Moving to that alternative is intellectually stimulating, it can inspire inquisitiveness, and it can encourage living in a thoughtful, responsible way. Geography is thus an invitation to life-long learning – one of the principal purposes of a liberal arts education.

Strengthening Civil Society and Policy-Making

As the foregoing chapters have shown, geography's perspective and analytical approach have much to contribute to the effort to confront many pressing political, social, economic, and environmental challenges. If geographical awareness is limited to a small set of sophisticated practitioners, however, the discipline's potential contributions to the policy-making process – and to making a better world – will be seriously constrained. Producing effective policies from the types of geographical understandings outlined in this book is unlikely to happen in the absence of a broad range of scientists, government officials, public intellectuals, and others who have been exposed to the discipline and who think geographically – people who are attuned to the importance of looking to what can be learned from a consideration of spatial patterns and place-based contextual differences; the interconnections across space that affect the development of places; nature–society dynamics; and the ways in which the geographical framing of issues affects how they are conceptualized. These are precisely the types of mental habits that geography education seeks to cultivate.

The memoir written in 1995 by Robert McNamara, US Secretary of Defense from 1961 to 1968, demonstrates their importance. McNamara attributes the disaster of Vietnam in part to the US policy elite's "profound ignorance of the history, culture, and politics" of the Vietnamese people.[5] The country's deepening engagement in Vietnam in the 1960s was driven by a Cold War metaphor of falling dominoes, which made the Communist orientation of the Ho Chi Minh regime the target of attention. What if greater attention had been directed to surging nationalism in Southeast Asia in the wake of a century of European colonialism and Japanese occupation (an issue that could hardly be missed when looking at the political geography of the region)? Might there have been some reconsideration of Vietnam's Cold War significance? Similarly, what if military strategists had been more willing to ask themselves what it was like to live in a village where grand geopolitical theories were meaningless, but where American soldiers were seen as the latest agents of death and destruction? Might there have been some reconsideration of the mission – and less surprise over the staunchness, perseverance, and tenacity of those who took up arms against soldiers they viewed as foreign invaders? No one can answer such counterfactual questions definitively, but they are suggestive of ways of thinking that geographical education encourages.

More recently we have seen tectonic shifts in the geopolitical landscape: upheavals in the Middle East and terrorism in the name of a variant of Islam, a resurgent Russia and a surging China, new attention to the Arctic in the wake of climate change, growing uncertainty about the strength of European unity, and abrupt political shifts in countries ranging from Pakistan to the Philippines. These developments are all rooted in concrete geographical circumstances, they are shaped by alliances and trading relationships that affect how individual places are connected to the rest of the world (i.e., geographical situation), and they play out against the backdrop of different views of how humanity should partition and use the surface of the planet (i.e., differences in geographical understanding). Efforts to grapple with these geopolitical shifts without the benefit of a geographical perspective – or even basic knowledge of geographical patterns – are inevitably compromised.

It follows that geographical literacy should not be thought of as a matter of importance solely for the governmental, policy, and scientific elite. A robust civil society depends on an informed, engaged general populace. In the absence of some grasp of the workings of the climate system and the major biophysical changes unfolding on Earth's surface, people are in a poor position to assess the validity of claims that a cold snap provides evidence that the climate is not warming. Without some awareness of North Korea's geographical situation in relation to its neighbors, it is impossible to assess the potential consequences of military and economic responses to the aggressive stance taken by the country's leaders. Widespread geographical ignorance means there is no check on misleading statements by public figures, the media, bloggers, and self-proclaimed pundits. (Think of the claims that attribute job losses in certain sectors to environmental regulations or immigrants, with no mention of competition from other places or consideration of how mechanization and transportation innovations have altered where and how goods are produced.)

Geographical education can make another signal contribution in the public arena. It can serve to promote mutual understanding. This claim might seem problematic given that geography once served the interests of colonial powers seeking to exert and maintain control over distant lands. In its modern guise, however, geography education invites consideration of what it means to see the world through the lens of those living in different places. That way of thinking makes it harder to vilify other people and places – a critical first step to conflict avoidance.

Facilitating Understanding and Use of Geospatial Technologies

The last few decades have played host to a technological revolution with deep geographical foundations. Most obviously, the use of GIS has become increasingly ubiquitous as a tool facilitating land-use planning, landscape architecture and building design, environmental assessment and management, the deployment of emergency services, spatially grounded academic research, and much more. GPS and online mapping sites such as Google Maps and Microsoft Virtual Earth have fundamentally changed the way people obtain directions and find their way to their destinations. Other online mapping applications make it possible for anyone with a computer to contribute geographical information to databases (so-called "Volunteered Geographic Information," or VGI[6]) – an activity that is helping to flesh out understandings of on-the-ground circumstances in remote places.

These developments point to another compelling justification for broad-based geographical education: it can help prepare people to understand the advantages and limitations of the geospatially infused technological environment of the twenty-first century. To start with GIS, given its burgeoning use, the job opportunities for individuals with skills in this area are great; it is thus not surprising that they are taught in many programs other than geography. Yet the quality and utility of a GIS analysis are determined by the types of spatial data chosen for the analysis, the judgments made about the weight to be given to each data layer, and decisions about the resolution of the data that go into each layer. Under the circumstances, the outputs of GIS analysis are not simply representations of the "real world"; they are (as discussed in chapter 2) the product of ideas and judgments that need to be constantly examined and assessed.

Thinking constructively and critically about GIS – its advantages and limitations – requires a good understanding of spatial data and analysis, a critical perspective on spatial frameworks, and a sensitivity to the ways in which choice of scale can influence analytical outcomes. These are all hallmarks of a good geographical education. They can, of course, be taught outside the formal geography curriculum, but they rarely receive the same degree of attention and emphasis. GIS analyses typically result in maps that can facilitate understanding of processes and options. Unfortunately, many such maps are difficult to interpret and are visually unappealing. In the hands of people with training in cartography and map design (long-standing core components of the geography curriculum), however, they can be tremendously useful, even influential.

Plate 9 is a map offering a clear, evocative picture of wildlife corridors in Wyoming. Coming out of a collaboration between geographers and wildlife biologists, the map draws attention to areas that are critical to the long-term health and survival of an important species. The map has been widely circulated and it was featured in a high-profile video produced in conjunction with the Wyoming Wildlife Initiative.[7] The map raised awareness of the importance of protecting wildlife corridors in the American West and it likely helped to lay the groundwork for one of the few pro-environment policies to come out of the early years of the Donald Trump administration in the United States: an Interior Department order calling for the study and preservation of habitat and migration corridors in Western states for big-game animals.[8] The order specifically calls for wildlife protection strategies that build on prior state migration initiatives, and the Wyoming initiative, backed by visualizations such as the one shown in plate 9, was the most visible and far-reaching of these. GIS practitioners who understand how to produce effective geographical visualizations are in a much better position to have this kind of influence.

Turning to GPS and online mapping platforms, no year goes by without stories of people following routes on platforms such as Google Maps, only to find themselves stuck, lost, or worse. There's the story from the United Kingdom of the Leicestershire woman who was driving to a christening some years ago and, after blindly following the prompt of her GPS, headed down a winding track and was swamped by a rising river. She managed to extricate herself from her expensive Mercedes, but her £96,000 automobile was lost.[9] On a more mundane level, the tendency blindly to follow Google Maps can lead an unsuspecting London cyclist pedaling east-to-west on a wet evening to head down a narrow, poorly lit, often congested towpath because that route is highlighted owing to its attractiveness to tourists on pleasant weekend afternoons.

In the absence of geographical education, it is easy to think of maps simply as depictions of the truth rather than as representations of temporally specific information that are subject to mistakes, just like any other human-produced information product. Maps of various kinds are, hearteningly, much more prevalent than they were a generation ago because they are much easier to produce and manipulate in a computer environment. Yet their very ubiquity raises the stakes for geography education. Film studies programs sprang up when producing and watching films became more common. Computer studies programs blossomed in the wake of the personal computer revolution. With maps now so much more a part of daily life, there is clearly an imperative to generate more widespread appreciation for cartography – not only for its value in facilitating communication, but also for its role in sensitizing people to the choices and biases that go into map making (as discussed in chapter 2).

Geography education can also counter one of the downsides of the widespread current practice of using GPS to move around cities, towns, and the countryside. While useful, GPS focus attention solely on a route and not on the overall organization of the landscape and the place of a route within it. Most GPS outputs show nothing about topography and they reveal nothing about where a given destination may be situated in relation to other places of possible interest. They thus discourage consideration of the surrounding context.[10] For all their geographical specificity, they can work against geographical understanding. Exposure to geography promotes awareness of these limitations and encourages efforts to overcome them (pulling out an old-fashioned road map, consulting an atlas, zooming out in an online environment, etc.).

In the wake of the development of online mapping platforms such as Wikimapia.org, MapAction.org, and OpenStreetMap.org, more and more non-professionals are contributing geographical information (VGI) that is filling important gaps in our understanding. Such efforts are already having a significant impact on disaster relief (helping emergency responders know where to go in the aftermath of earthquakes), humanitarian aid (providing information about refugee patterns of movement and places where aid packages are particularly needed), and public health monitoring (facilitating fast reporting of the precise location of disease outbreaks). The more people are exposed to geography and geospatial technologies, the more they are likely to want to contribute further to these kinds of efforts.

Despite these potentials of geospatial technologies, they also raise significant privacy questions. Most people with access to computer technology and credit cards leave a rich trail of geocoded information that allows government agencies and marketing firms to build vast databases of personal information. People can be tracked without their knowledge, they can be targets of advertising for unwanted products, and compromising information about their activity patterns can be used against them. Coming to grips with the consequences of this state of affairs – and designing encryption protocols that can reduce the risks of abuse – will require a better understanding of where the most significant threats to human privacy are found, and how geospatial technologies can be configured to protect sensitive information. Once again, the importance of geography education looms large, as it will take a range of people steeped in geographical ways of thinking and the workings of geospatial technologies to confront these issues.

Conclusion

Education serves many purposes, including imparting knowledge and skills that are needed to move society forward, enabling students to adapt to a world that will change during the course of their lifetimes, and bringing more meaning to people's lives. Geography has important contributions to make to all of these ends. It offers people critical insights into the organization and character of the world around them, and it allows individuals to understand technologies that are affecting their lives. Geography yields insights that can help students of the discipline understand the changes unfolding around them and learn how to use tools for assessing and adapting to those changes. Moreover, it opens people's eyes and minds to the richness and wonder of the surrounding world; it heightens awareness of – and by extension concern for – distinctive places and environments; and it fosters curiosity that is rewarding in its own right. Geography is, in short, a key to making sense of our increasingly connected, crowded, environmentally fragile, and rapidly changing world.

Coda

The past 2,000 years has seen major changes in the geography of the planet. Volcanic eruptions virtually erased some islands from the map, even as they added substantially to the landmass of others. A medieval warm period altered the biogeography of Europe and made it possible to cultivate crops in more northern latitudes. Innovations in ship design led to a massive exchange of peoples and goods between the western and eastern hemispheres, as well as the decimation of a significant portion of the population of the former. The invention of the railroad, and later the automobile, led to enormous demographic shifts, transformed the size and layout of urban areas, and changed the spatial organization of production and consumption.

There is much to be learned from studying the geographical changes of the past two millennia that gave rise to the contemporary world, but we cannot stop there. Many of those developments unfolded across many decades, if not centuries. Now we are living in a world experiencing major changes over the course of much shorter time periods, and it seems all but certain that the pace of change will accelerate in the years and decades ahead. The environment is being remade before our eyes, the geopolitical map is in great flux, cities are exploding, the connections between places and peoples are being remade, and the development and diffusion of new technologies is rapidly altering how we live, how we relate to one another, and even how we think about ourselves and our environment.

It will take a massive effort on many different fronts to comprehend, much less constructively address, the changes that are taking place today. But geography must surely be an important part of that effort, for it is the geography of the planet itself that is being transformed. We are awash in data about the transformations taking place, but we cannot hope to gain a handle on them if the population at large has little sense of Earth's geographical character and the changes that are happening to it; if students and scholars lack the analytical perspectives and tools needed to assess the evolving spatial organization and material character of places and regions; or if policy-makers and planners are not equipped to think geographically about issues and problems – to think knowledgeably and critically about geographical patterns, to consider why things happen where they do, and to appreciate how geographical context influences what happens.

Consider how important geographical thinking and understanding will be to the effort to confront just one facet of the mobility revolution that was touched on at the end of chapter 2: the replacement of most conventional cars and trucks by connected, automated, shared, electric (CASE) vehicles within the next quarter-century. We are on the cusp of seeing streets filled with shared vehicles that are guided on digital tracks in narrower lanes than are currently in use, moving down streets that have no curbs. If the predictions of the technology sector are even remotely accurate, it will be increasingly uneconomical to own individual automobiles, and newer vehicles will be safer and easier to maintain than the conventional vehicles of today.

There are obvious first-order knock-on effects of this transformation. The character of streets and sidewalks will change. Most gas stations will disappear. The automobile industry will undergo a radical makeover. The oil industry will shrink dramatically. Yet if we stop there (and unfortunately much of the discussion of CASE does stop there), we will have a very limited understanding of what lies ahead.

A 2017 American Geographical Society symposium at Columbia University on the future of mobility drew attention to the broader-ranging, longer-term consequences of CASE and related developments: changes in the organization of public transportation, land use, municipal finances, air quality, employment opportunities, urban growth patterns, place-to-place interconnections, and more. They also have the potential to alter fundamentally the spatial organization of cities, affecting patterns of wealth and poverty, the organization of economic activities, the demographic and ethnic composition of neighborhoods, people's daily activity patterns, and even their sense of place. Given that each of these is fundamentally geographical in nature, the kinds of geographical perspectives and tools discussed in this book will be essential to understanding them and shaping their development in constructive ways.

Grasping the nature of the changes that are coming, much less anticipating them, will require a sensitivity to the evolving geography of the planet, as well as the capacity to think geographically. We simply cannot afford to stumble through our lives wearing geographical blinders – giving only limited attention to how people, environments, and places are organized and interconnected, or failing to nurture the ability to think carefully and critically about where things happen, why they happen where they do, and how geographical context influences environmental processes and human affairs. These concerns are sufficiently fundamental to navigating the twenty-first century that an understanding and appreciation of geography should not be seen as a luxury; it should be viewed as vital to the effort to create a more livable, just, sustainable, and peaceful planet.

Notes

Chapter 1 Geography's Nature and Perspectives

1 See generally Ben Taub, "Lake Chad: The World's Most Complex Humanitarian Disaster," *New Yorker*, December 4, 2017. Available at https://www. newyorker.com/magazine/2017/12/04/lake-chad-the-worlds-most-complex-humanitarian-disaster.

2 National Research Council, *Understanding the Changing Planet: Strategic Directions for the Geographic Sciences* (Washington, DC: National Academies Press, 2010), p. ix.

3 Like most fields of study, there is no easy, universally accepted, definition of geography. Shorthand efforts to define it include "the study of spaces and places on Earth's surface," "the why of where," and "the study of the difference that space makes in human and bio-physical processes." Each of these has its advantages and limitations.

4 There is some uncertainty as to the origins of the "why of where" expression, but it probably was

coined by Marvin Mikesell, a geography pr
the University of Chicago from 1958 to 201/.

5 Stuart Elden, "Reassessing Kant's Geography," *Journal of Historical Geography*, 35:1 (2009): 3–25, quote p. 14.

6 R. D. Dikshit, *Geographical Thought: A Contextual History of Ideas* (Delhi: Prentice-Hall of India, 1997), pp. 3–4.

7 National Research Council, *Rediscovering Geography: New Relevance for Science and Society* (Washington, DC: National Academies Press, 1997).

8 The colleague was Ronald Wixman, who taught geography at the University of Oregon from 1975 to 2006.

Chapter 2 Spaces

1 Peter Jordan, "The Problems of Creating a Stable Political-Territorial Structure in Hitherto Yugoslavia," in Ivan Crkvenčić, Mladen Klemenčić, and Dragutin Feletar, eds., *Croatia: A New European State* (Zagreb: Urednici, 1993): 133–42.

2 Thomas Friedman, *The World is Flat: A Brief History of the Twenty-First Century* (New York: Farrar, Straus and Giroux, 2005).

3 Laurence Smith, Yongwei Sheng, and Glen MacDonald, "Disappearing Arctic Lakes," *Science*, 308 (2005): 5727.

4 Anthony Clavane, "Brexit Heartland and City of Culture Hull Remains in Dangerous Waters," *The*

New European, December 19, 2017. Available at http://www.theneweuropean.co.uk/top-stories/brexit-heartland-and-city-of-culture-hull-remains-in-dangerous-waters-1-5322162.

5 Neil M. Coe, Martin Hess, Henry Wai-chung Yeung, Peter Dicken, and Jeffrey Henderson, "'Globalizing' Regional Development: A Global Production Networks Perspective," *Transactions of the Institute of British Geographers*, 29:4 (2004): 468–84.

6 See generally Alexander B. Murphy, "The Sovereign State System as Political-Territorial Ideal: Historical and Contemporary Considerations," in Thomas Biersteker and Cynthia Weber, eds., *State Sovereignty as Social Construct* (Cambridge: Cambridge University Press, 1996): 81–120.

Chapter 3 Places

1 Kimberly Lanegran and David Lanegran, "South Africa's National Housing Subsidy Program and Apartheid's Urban Legacy," *Urban Geography*, 22:7 (2001): 671–86.

2 Ibid.

3 Max Moritz, Chris Topik, Craig Allen, Tom Veblen, and Paul Hessburg, "SNAPP Team: Fire Research Consensus" (Collaborative Project of the Nature Conservancy, the Wildlife Conservation Society, and the National Center for Ecological Analysis and Synthesis at the University of California, Santa Barbara). Available at https://snappartnership.net/teams/fire-research-consensus/.

4 Jonathan D. Phillips, "Human Impacts on the Environment: Unpredictability and the Primacy of Place," *Physical Geography*, 22:4 (2001): 321–32.

5 Linda McDowell and Doreen Massey, "A Woman's Place?" in Doreen Massey and John Allen, eds., *Geography Matters!* (Cambridge: Cambridge University Press, 1984): 128–47.

6 Food and Agriculture Organization of the United Nations, *Voluntary Guidelines for Securing Sustainable Small-Scale Fisheries in the Context of Food Security and Poverty Eradication* (Rome: Food and Agriculture Organization, 2015).

7 National Academies of Sciences, Engineering and Medicine, *Communities in Action: Pathways to Health Equity* (Washington, DC: National Academies Press, 2017).

8 Carlos A. Nobre, Gilvan Sampaio, Laura S. Borma, Juan Carlos Castilla-Rubio, José S. Silva, and Manoel Cardoso, "Land-Use and Climate Change Risks in the Amazon and the Need of a Novel Sustainable Development Paradigm," *Proceedings of the National Academy of Sciences of the United States of America*, 113:39 (2016): 10759–68.

9 This question was taken up in Michael K. Reilly, Margaret P. O'Mara, and Karen C. Seto, "From Bangalore to the Bay Area: Comparing Transportation and Activity Accessibility as Drivers of Urban Growth," *Landscape and Urban Planning*, 92:1 (2009): 24–33.

10 Edward Relph, *Place and Placelessness* (London: Pion, 1976).

11 Kay J. Anderson, "The Idea of Chinatown: The Power of Place and Institutional Practice in the Making of a Racial Category," *Annals of the Association of American Geographers*, 77:4 (1987): 580–98.

12 Samuel P. Huntington, *The Clash of Civilizations and the Remaking of World Order* (New York: Simon & Schuster, 1997).

13 Stephen Zunes, "Bush Again Resorts to Fear-Mongering to Justify Iraq Policy," *Foreign Policy in Focus* (October 12, 2005). Available at https://fpif.org/bush_again_resorts_to_fear-mongering_to_jus tify_iraq_policy/.

14 Nick Halley, *Terrorism: The Target is You! The War Against Radical Islam* (Self-Published, 2004).

15 Linda Lobao, Gregory Hooks, and Ann Tickamyer, eds., *The Sociology of Spatial Inequality* (Albany: State University of New York Press, 2007).

Chapter 4 Nature and Society

1 Tim Dee, "Our Bleak Exile of Nature," *New York Times*, May 1, 2015. Available at https://www.nytimes.com/2015/05/02/opinion/our-bleak-exile-of-nature.html.

2 See, e.g., Robert D. Kaplan, *The Revenge of Geography: What the Map Tells Us About Coming Conflicts and the Battle Against Fate* (New York: Random House, 2012), and David S. Landes, *The Wealth and Poverty of Nations: Why Some Are So Rich and Some So Poor* (New York: W. W. Norton & Company, 1999).

3 K. O'Brien, R. Leichenko, U. Kelkar, H. Venema, G. Aandahl, H. Tompkins, A. Javed, S. Bhadwal, S. Barg, L. Nygaard, and J. West, "Mapping Vulnerability to Multiple Stressors: Climate Change and Globalization in India," *Global Environmental Change*, 14:4 (2004): 303–13.

4 See, e.g., Laura Pulido, "Rethinking Environmental Racism: White Privilege and Urban Development in Southern California," *Annals of the Association of American Geographers*, 90:1 (2000): 12–40.

5 B. L. Turner, Eric F. Landin, and Anette Reenberg, "The Emergence of Land Change Science for Global Environmental Change and Sustainability," *Proceedings of the National Academy of Sciences of the United States of America*, 104:52 (2007): 20666–71.

6 United States Geological Survey, *Land Change Science Program* (website, December 2, 2016). Available at https://www2.usgs.gov/climate_landuse/lcs/.

7 Gilbert F. White, *Human Adjustment to Floods: A Geographical Approach to the Flood Problem in the United States* (Chicago: University of Chicago Geography Research Series no. 29, 1945).

8 Piers Blaikie, *The Political Economy of Soil Erosion in Developing Countries* (Abingdon, Oxon: Longman Scientific and Technical, 1985).

9 Diana M. Liverman, "Drought Impacts in Mexico: Climate, Agriculture, Technology, and Land Tenure in Sonora and Puebla," *Annals of the Association of American Geographers*, 80:1 (1990): 49–72.

10 See, e.g., Hallie Eakin, Alexandra Winkels, and Jan Sendzimir, "Nested Vulnerability: Exploring Cross-Scale Linkages and Vulnerability Teleconnections in Mexican and Vietnamese Coffee Systems," *Environmental Science & Policy*, 12:4 (2009): 398–412.

11 John O'Loughlin, Frank Witmer, Andrew Linke, Arlene Laing, Andrew Gettelman, and Jimy Dudhia, "Climate Variability and Conflict Risk in East Africa, 1990–2009," *Proceedings of the National Academy of Sciences of the United States of America*, 109:45 (2012): 18344–9.

12 For a general discussion of this issue, see William G. Moseley, Eric Perramond, Holly M. Hapke, and Paul Laris, *An Introduction to Human–Environment Geography: Local Dynamics and Global Processes* (Hoboken, NJ: John Wiley & Sons, 2013).

13 Joshua Muldavin, "From Rural Transformation to Global Integration: Comparative Analyses of the Environmental Conditions of China's Rise," *Eurasian Geography and Economics*, 54:3 (2013): 259–79.

14 William G. Moseley, "Farmers in Developing World Hurt by 'Eat Local' Philosophy in US," *San Francisco Chronicle*, November 18, 2007. Available at https://www.sfgate.com/opinion/article/Farmers-in-developing-world-hurt-by-eat-local-3301224.php.

15 For a critique of the "Global South" moniker, see Alexander B. Murphy, "Advancing Geographical Understanding: Why Engaging Grand Regional Narratives Matters," *Dialogues in Human Geography* 3:2 (2013): 131–49.

16 The discussion draws on Leslie McLees, "Intersections and Material Flow on Open-Space Farms in Dar es Salaam, Tanzania," in Antoinette WinklerPrins, ed., *Global Urban Agriculture: Convergence of Theory and Practice between North and South* (Wallingford, UK: CABI International, 2017): 146–58.

17 Quoted in Michael Hobbes, "Stop Trying to Save the World," *New Republic*, November 17, 2014. Available at https://newrepublic.com/article/120178/problem-international-development-and-plan-fix-it.

Chapter 5 Why We All Need Geography

1 Roper Public Affairs and National Geographic, *2006 Geographic Literacy Study* (New York: GfK NOP, 2006). Available at https://media.national geographic.org/assets/file/NGS-Roper-2006-Report.pdf.

2 On this point see Daniel Hallin, "Whatever Happened to the News?" (Center for Media Literacy, n.d.). Available at http://www.medialit.org/reading-room/whatever-happened-news.

3 The cartoon by Gary Varvel appeared on July 23, 2007. Available at http://www.cartoonistgroup.com/store/add.php?iid=19562.

4 See, e.g., Selin Kesebir and Pelin Kesebir, "How Modern Life Became Disconnected from Nature," *Greater Good Magazine*, September 20, 2017. Available at https://greatergood.berkeley.edu/article/item/how_modernlife_became_disconnected_from_nature.

5 Robert S. McNamara, *In Retrospect: The Tragedy and Lessons of Vietnam* (New York: Times Books, 1995).

6 The term was coined by Michael Goodchild to describe private citizen involvement in creating, assembling, and disseminating geographical information. Michael Goodchild, "Citizens as Sensors: The World of Volunteered Geography," *GeoJournal*, 69:4 (2007): 211–21.

7 The video can be found on the website of the Wyoming Wildlife Initiative at http://migrationinitiative.org/content/red-desert-hoback-migration-assessment.

8 US Department of the Interior, Order number 3362 (February 9, 2018). Available at https://www.doi.gov/sites/doi.gov/files/uploads/so_3362_migration.pdf.

9 Andy Dolan, "£96,000 Merc Written Off as Satnav Leads Woman Astray," *Daily Mail*, March 16, 2007. Available at http://www.dailymail.co.uk/news/article-442730/96-000-Merc-written-satnav-leads-woman-astray.html.

10 John Edward Huth, *The Lost Art of Finding Our Way* (Cambridge, MA: Harvard University Press, 2013).

Further Reading

It is difficult to capture geography's breadth and vibrancy in a short "further reading" list, but the following entries will give general, non-specialist readers a deepened appreciation of what it means to look at the world through a geographical lens. The entries are organized into three broad categories: general books on geography written by professional geographers, books by geographers that wrestle with particular geographical topics and themes, and books by non-geographers that reflect geographical thinking.

General Books on the Nature of Geography Written by Geographers

Danny Dorling and Carl Lee, *Geography* (London: Profile Books, 2016).

Written by two distinguished British geographers, this book looks at how geography reflects and shapes globalization, inequality, and sustainability. With additional commentary on the discipline's past traditions and what it can tell us about the future, the book offers a valuable overview of geography's essential character and ambition.

Susan Hanson, ed., *Ten Geographic Ideas that Changed the World* (New Brunswick, NJ: Rutgers University Press, 1997).

Rather than looking broadly across geography, this book devotes a chapter to each of ten influential geographical concepts and practices. Engaging essays on topics such as the map, sense of place, and human adjustment demonstrate geography's contributions to human understanding.

National Research Council, *Rediscovering Geography: New Relevance for Science and Society* (Washington, DC: National Academies Press, 1997).

Produced in the wake of growing recognition of geography's importance, this book offers an informative overview of the types of thinking, research, and tools geography brings to the study of critical issues facing society. It helped to raise geography's profile in the United States and set the stage for follow-on National Academies studies (see, e.g., chapter 1, note 2).

Further Reading

Studies by Geographers on Particular Topics and Themes

Harm J. de Blij, *Why Geography Matters More Than Ever* (Oxford: Oxford University Press, 2012).

In lively, readable prose, Harm de Blij makes the case that geography offers an important window on global issues ranging from climate change to the rise of China to upheavals in the Middle East. The book also underscores how geographical situation continues to affect the lives and fates of people around the world – a central theme in one of his prior books (*The Power of Place: Geography, Destiny, and Globalization's Rough Landscape* [Oxford: Oxford University Press, 2008]) that stands as a counterpoint to Thomas Friedman's flat-world postulate (see chapter 2, note 2).

Mona Domosh and Joni Seager, *Putting Women in Place: Feminist Geographers Make Sense of the World* (New York: Guilford Press, 2001).

This insightful book opens up thinking about the myriad ways in which gendered thinking and practices shape geographical arrangements and understandings. The book helped move geography away from its long male-dominated orientation, and its broad historical and geographical sweep make it a volume of continuing relevance.

Further Reading

Andrew Goudie and Heather Viles, *Landscapes and Geomorphology: A Very Short Introduction* (Oxford: Oxford University Press, 2010).

Distinguished physical geographers Goudie and Viles offer a lively, informative overview of the evolution of the physical landscape – extending beyond Earth's land surface to the ocean floors, Mars, and Titan. Their focus on the role of intersecting forces in landscape change – geologic, climatologic, and human – shows what it means to look at the physical world through a geographical lens.

Martin W. Lewis and Kären Wigen, *The Myth of Continents: A Critique of Metageography* (Berkeley: University of California Press, 1997).

Lewis and Wigen expose the taken-for-granted divisions that are often used to make sense of the world. They challenge readers to think carefully and critically about the geographical assumptions underlying the ways they divide up the world – in the process showing why careful geographical thinking is so important.

Mark Monmonier, *How to Lie with Maps*, 2nd ed. (Chicago: University of Chicago Press, 2014).

An update of a classic book, Monmonier offers an engaging, insightful assessment of the use and abuse of maps. The book challenges readers to treat maps as the product of ideas, perspectives, and prejudices, not simply as objective representations of reality.

Laurence C. Smith, *The World in 2050: Four Forces Shaping Civilization's Northern Future* (New York: Dutton, 2010); published in the United Kingdom as *The New North: The World in 2050* (London: Profile Books, 2012).

Using the tools and techniques of geography to look forward in time, Smith considers how demographic, environmental, and resource issues will shape the Arctic (and other places) in the decades to come. The book provides a splendid example of how the integrative instincts of a geographer specializing in hydrology, glaciology, and remote sensing can lead to broader, more synthetic thinking about our planetary home.

Yi-Fu Tuan, *Space and Place: The Perspective of Experience* (Minneapolis: University of Minnesota Press, 1977).

A seminal contribution that promoted awareness of geography's humanistic dimension, Tuan artfully shows that geography's two core concerns – space and place – are not simply phenomena to be modeled and described abstractly. They are fundamental to the human experience and need to be understood as such.

Further Reading

Geographical Studies by Non-Geographers

Jared Diamond, *Guns, Germs, and Steel: The Fates of Human Societies* (New York: W. W. Norton & Co., 1997).

A remarkable, if controversial, effort to show that the comparative success of different civilizations resulted from variations in geographical context rather than intellectual or moral differences.

David R. Montgomery, *Dirt: The Erosion of Civilizations* (Berkeley: University of California Press, 2012).

A sweeping overview of the relationship between soil and civilizations, showing how the latter have used, and misused, soil in ways that have degraded one of Earth's most important natural endowments.

Saskia Sassen, *The Global City: New York, London, Tokyo*, 2nd ed. (Princeton: Princeton University Press, 2001).

A geographically sensitive discussion of how networks, financial flows, and labor mobility are shaping the development of global cities, with consequences for urban form, social stability, and sustainability.

Further Reading

Andrea Wulf, *The Invention of Nature: Alexander von Humboldt's New World* (New York: Knopf, 2015).

A penetrating account of the life and work of one of the progenitors of the modern discipline of geography.

Index

Index

Index

drought 3, 7, 14, 92–4, 99–101
  see also precipitation
drugs 51

East Asia 7, 56
ecologists 11–12, 91
ecology 2, 5, 26, 65, 70, 87, 91, 98–9
economic activities 10, 20, 124, 136
economic circumstances 4–5, 7, 41, 43, 46, 50, 70, 77–8, 98–100, 105, 121
economic geography 27, 32, 39, 48, 93, 119
economic information 26
ecosystem 1, 35, 91, 94, 96
education 20, 118–21
  geography 8, 13–15, 20–2, 27–30, 90, 95, 110–12, 118–21, 123–6, 128–9, 131
empires 16–19, 79, 84
Enlightenment 18
environmental alteration 2, 4–5, 9, 57–9, 103, 109, 134
  see also human–environment relations
environmental context 1, 8, 18, 25, 45, 69, 87, 89, 101, 114, 116–19, 132, 137
environmental determinism 90, 105
environmental geography 13, 15, 20, 23, 40–1
environmental justice 96

environmental management 124–5, 127
environmental problems 9, 39, 45, 52, 58
environmental processes 23
environmental sustainability 36, 100, 103, 106–7
erosion 37, 41, 98, 102
ethnic conflict 5–6
ethnic patterns 14, 15, 25, 35, 37–8, 54, 80, 114–15, 136
ethno-cultural groups 1, 78
ethnographic methods 29
Europe 4, 7, 18, 23, 25–6, 50–3, 56, 60, 79, 87, 103, 114, 133
European Union 46, 113, 122

feminist geography 27
  see also gender
field trips 88, 118
financial flows 40
fire 45, 65–7
fisheries 68–9
fishing 2, 46, 51, 68–9
floodplains 97–8
floods 17, 43, 45, 50, 92, 94, 119
food 1–3, 31, 37, 103, 116
forests 15, 26, 35, 37, 45, 61, 65–6
France 7, 22, 48, 53, 75, 79
Friedman, Thomas 41–3

gender 27, 67, 69
  see also feminist geography
gentrification 15
geoarcheology 12

Index

Index

Index

Index

Index

Index

watersheds 50
Watkins, Derek 115
White, Gilbert 97–8
Wikimapia.org 130
wildlife 28, 127

Wyoming Wildlife Initiative
127–8

Yangtze River 50
Yugoslavia 54